Science Web Reader

BIOLOGY

Author team:

Joan Solomon (Series Editor): Visiting professor at the Open University and King's College London.
An experienced author who specialises in history of science and ethics and has written extensively at KS3/KS4.

Peter Horsfall: Senior lecturer in science education at John Moores University Liverpool.
Author of Go for Science.

Michael Reiss: Reader in Education and Bioethics at Homerton College.
Author of a number of very successful biology texts.

Pat O'Brien: Educational consultant and former science adviser. Series editor for Challenge Science.

Richard Robinson: Television presenter and writer. Author of a series of books called Science Magic.

Ruth Hughes and **Chris Dunning** are both science editors.

Acknowledgements

BSIP Gems/Europe/Science Photo Library **15b**; Bettmann/Corbis UK Ltd. **15t**; Corbis UK Ltd. **16**;
Martin Jones/Corbis UK Ltd. **28b**; *Michael Maconachie*; Papilio/Corbis UK Ltd. **28c**;
Dr Jeremy Burgess/Science Photo Library **28t**; *David Guyon*, The BOC Group Plc/Science Photo Library **30**;
NASA/Science Photo Library **34**; *Ray Simons*/Science Photo Library **36c**; Biophoto Associates **36l**;
David Scharf/Science Photo Library **36r**; Vatican Museum and Galleries/Bridgeman Art Library **38**;
Lester V. Bergman/Corbis UK Ltd. **40c**; Biophoto Associates **40l**; *Lester V. Bergman*/Corbis UK Ltd. **40r**;
George McCarthy/Corbis UK Ltd. **51**; *Maurizio Lanini*/Corbis UK Ltd. **54**; *Roger Antrobus*/Corbis UK Ltd. **55**;
Karl Ammann/Corbis UK Ltd. **56**; *D. Robert Franz*/Corbis UK Ltd. **58**; Biophoto Associates **5b**;
The Stapleton Collection/Bridgeman Art Library **5t**; *Jack Fields*/Corbis UK Ltd. **60**; *B.J.W. Heath* **63l**;
Heather Angel/Natural Visions **63r**; *Ian Tait*/Natural Visions **64l**; Oxford Scientific Films **64r**;
Science Photo Library **6b**; Science Photo Library **6t**; Biophoto Associates **7**; Bettmann/Corbis UK Ltd. **9**.

First published in 2000
Nelson
Delta Place
27 Bath Road
Cheltenham
GL53 7TH
United Kingdom

A catalogue record for this book is available from the British Library.

ISBN 0-17-438737-7

00 01 02 03 04 / 10 9 8 7 6 5 4 3 2 1

Printed in Croatia by Zrinski d. d. Cakovec

CONTENTS

Foreword

These readers for Science Web are a new departure for science education. They have been written for students of all sorts, for science enthusiasts who want to read more about modern science, for students who are on the humanities side and who like nothing more than curling up with a good story – whether about a scientist or anyone else, and for students who don't usually read much at all but enjoy cartoons and good illustrations. Some of the stories are about exciting new scientific discoveries, some are about scientific explorations in earlier times and a few describe applications of science or technology. They aim to keep school science up to date and interesting.

Teachers will find that the subject matter of each reading passage connects with science in the National Curriculum, even though the stories do not try to teach it in the usual way. Because the stories are related to real people we believe their science content will become more memorable. At the end of each article there are a selection of questions. These might be used by some teachers as part of the homework, or for class work when students seem to need a quiet spell of self-learning. Other teachers may choose to adapt or even ignore the questions.

We believe that all the articles are suitable for some Key Stage 3 readers. We know that students in these years of schooling have a wide range of reading ages and rather than reduce the language to some lowest common denominator we decided to replicate the natural variety in two ways. Firstly we have provided stories with different styles, some deliberately exciting, some amusing, and some which pursue the search for a solution to some medical or similar problem. Our second way of providing variety is through the illustrations. Our artists have produced exceptionally fine pictures, some of which are quite beautiful. There are also photographs and cartoons.

These readers are designed to spread enthusiasm for science. It is changing so fast that we need to include up-to-the-minute discoveries. The stories from earlier times show that science has always been changing. It is, and was, the product of inventive men and women, so we have included human detail of how different scientists have reacted to challenges. Most importantly, these stories are designed to encourage students to get into the habit of reading about science, and so kindle a lifelong interest in it and in its progress.

Professor Joan Solomon

Series Editor

Science Web Readers

Microscopes and cells

Observation is an important skill for scientists. Many new discoveries are made after careful observation of things or events. So you can imagine how exciting it is for scientists when a new invention makes it possible to observe the world in a much more detailed way. This is what happened when the first microscopes where invented. They allowed us to look closely at the structure of living things for the first time. The seventeenth century was a great time for careful observation.

In fact, glass lenses had been around for hundreds of years. They were used by the ancient Chinese, the Greeks, and the Romans – but they were mainly used to start fires. It took a long time for us to realise that they could also be used to magnify things. It was not until the fourteenth century that the Italians began to use them in eye-glasses. What a change this made to people's lives! Soon, lens making became a major industry in Europe, with its centre in Holland.

The first microscopists

Zaecharias Jesen was the son of a Dutch spectacle maker – his father made lenses to fit spectacles. These were sold in his shop, where customers would come and try out lenses to find a pair with which they could see more clearly. Zaecharias did experiments with the lenses that his father made. In about 1600, he discovered that putting one lens behind the other made objects look many times bigger than using one lens alone. He had invented the compound microscope.

In 1609, Galileo Galilei, the great Italian scientist, made a better microscope by fitting the two lenses into a tube. He could adjust the tube to move the lenses to help him focus on an object. Galileo said that his microscope 'makes a flea look the size of a hen'! He made many observations of animals and plants with his microscope. Unfortunately, perhaps, Galileo also invented a telescope, which made observing the moon and planets easier. Because of the importance of these discoveries, people would be less interested in using microscopes until some years later.

Antony van Leeuwenhoek (1632–1723) was the son of a basket maker in the town of Delft in Holland. He grew up to become a shopkeeper and council official. In his spare time he had a hobby – lens making. He was passionate about his hobby and would spend many hours making observations with magnifying glasses.

Although microscopes had been recently invented, Antony was not impressed with them. They were difficult to keep in focus. He learned to grind his own lenses, and used them to make small microscopes.

A spectacle shop, from around 1600.

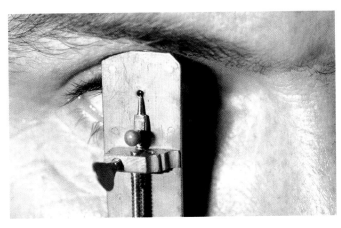

Antony van Leeuwenhoek's microscope.

5

Antony was devoted to lens making. He made over 400 lenses in his lifetime and gave many away to his friends. But he always kept the best ones for himself. They could magnify things up to 240 times.

Antony observed a wide variety of things using his simple microscopes. He was a careful observer, and left a wealth of records of his observations. After all, he was the first human to see the things he observed! He saw the microscopic life that lives in ponds, and the bacteria that grow on our teeth. He was the first to see the tiny blood vessels – capillaries – in the skin of a tadpole's tail, and the spermatozoa, or male sex cells, in human semen.

He wrote to scientists of the Royal Society in London about his microscopes and his observations. In his will, he left 26 microscopes to the Society.

Great discoveries

Antony's love of microscopes soon spread to other scientists. At this time, Robert Hooke was the curator of experiments at the Royal Society. He had a fascination for the many new pieces of scientific equipment that were being developed at the time, particularly the compound microscope.

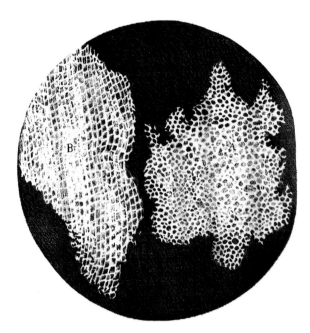

Cork under a microscope. This was the first picture of 'cells' in a biological specimen, named as such by Robert Hooke.

Like others, he observed all sorts of things using this new invention. He produced a book of his drawings in 1665, including the first ever drawing of cells, which he had seen in thin slices of cork. He wrote about his observations, giving clear descriptions of what he had done and seen:

'*I took a good clear piece of Cork, and with a Pen-knife sharpened as keen as a Razor, I cut an exceeding thin piece of it, and placing it on a black object Plate because it was itself a white Body, and casting the light on it ... I could exceeding plainly perceive it to be all perforated and porous much like a Honeycomb in these particulars.*

First in that it had a very little solid substance, in comparison of the empty cavity that was contained between.

Next, in that these pores, or cells, were not very deep, but consisted of a great many little Boxes, separated out of one continuous long pore by certain Diaphragms, as is visible in the Figure B which represents a sight of these pores split the long ways.'

Robert Hooke's drawing of his own compound microscope in Micrographica *(1665).*

Hooke was the first to record observations of cells. He saw these 'little boxes' in many different plants, including carrots, ferns and reeds, but he did not realise how important they were. He never recorded seeing what was inside cells. He believed that they simply provided a framework inside plants.

At about the same time, in 1665, Jan Swammerdam was looking at blood through a microscope. Jan was a Dutch scientist who had been bitten by the microscope bug. He would often work from dawn until dusk using lenses to magnify small insects and other animals, such as snails, while he cut them up to see what they were like inside. In blood he saw 'a vast number of roundish particles of flat, oval but regular form'. This was the first description of red blood cells.

The microscope craze

During the eighteenth century, looking through lenses was a craze as big as playing computer games is today! Everybody who could afford one wanted a microscope. People would have microscope parties, at which they would spend the evening showing each other what things looked liked through the instrument. Anything and everything was prepared for observation – pieces of hair, cotton, butterflies' wings, spiders' legs.

Although they were popular, microscopes became less important in making scientific discoveries for a while. They were very difficult to use, and could not easily be focussed. No sooner were they focussed than coloured fringes would appear around the image of the object being examined. This made it hard to make accurate observations.

Better microscopes

Lens makers eventually found a solution to this problem. They had to use different kinds of glass for each of the lenses in the microscope. These were called achromatic microscopes, and were invented in 1830. At the same time, a machine was invented that allowed scientists to cut thin slices of material to look at through the microscope.

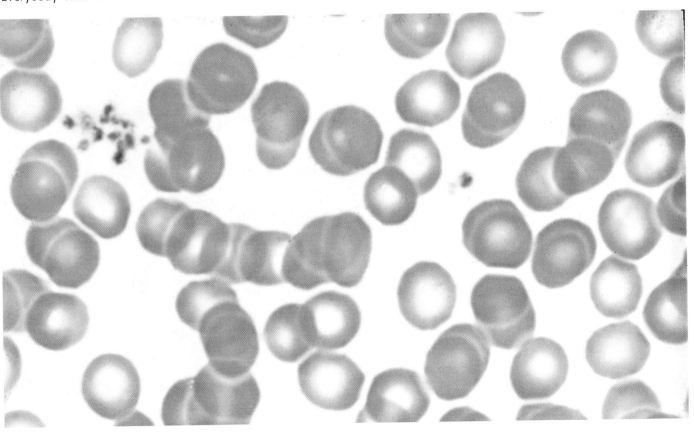

Red blood cells under the microscope.

In 1833, Robert Brown, a Scottish scientist, was the first to describe his observation of the nucleus in plant cells. He imagined that the nucleus might play an important part in the life of the cell.

One hundred and seventy years after Robert Hooke first described cells, someone found that cells contained living material. In 1835, Felix Dujardin, a French scientist, recognised that cells contained a substance that he called 'sarcode' – what we call cytoplasm today.

Henri Dutrochet, in 1824, suggested that cells are alive, and that chemical processes go on inside them. He saw the chloroplasts in plant cells, and believed they were important in how plants used carbon dioxide.

The final realisation of how important cells were in all living things was soon to come. In 1838, a German called Mathias Schleiden announced that, as far as he could tell, all plants were made up of cells. The next year, Theodore Schwann made the same announcement for animals!

Over the next forty years, more efforts were made to make studying cells easier. New methods were developed for preparing slides so that cells could be seen more clearly. Coloured stains were developed that showed us what was inside cells. In 1870, chromosomes were discovered in the nucleus. Scientists began to understand how information passes from one generation to the next.

Our understanding of cells has grown because of the work of scientists from many different countries over a period of two hundred years. As with so many areas of human endeavour, each scientist has made a contribution, building on the work of others who went before.

QUESTIONS

1. Draw a time line to show how lenses and microscopes developed from the time of the ancient Romans.
2. Draw a time line to show the important observations in our understanding of cells.
3. Carefully read Robert Hooke's description of how he observed 'cells' in cork.
a) Rewrite the passage, making it easier for a modern reader to understand.
b) Write a set of instructions to tell a friend how to observe cork using a microscope.
4. Look at some pictures of cells. Write an accurate description of one of them, but without using a diagram. Ask a friend to draw a cell, using only the information you have given them.
5. Write a letter to Antony van Leeuwenhoek. Tell him about what has been discovered about cells since he first used his simple microscope. Ask him to explain how important his own discoveries were.
6. Look at the drawings of early microscopes. Write a set of instructions explaining how you think they were used.

The man with a lid on his stomach

Many of the greatest scientific discoveries have come about as a result of a clever person being in the right place at the right time. This was true for Dr William Beaumont, the first person to show by experiment how food is digested in the stomach.

Scientists had wondered for hundreds of years about what happens to food inside the body. They knew that food was needed to provide important substances to keep the body alive and healthy. They also knew that it was broken down into a soup-like liquid inside the stomach and intestine. But they did not understand how this happened. Scientists had different ideas to explain the process:

- The ancient Greeks believed that food was broken down by heat in the stomach, in a process similar to cooking.
- Other early scientists believed that food simply broke down by rotting away inside the stomach – although some knew that the acid in the stomach prevented the rotting of food.

Then, in the 18th century, the Italian scientist Lazaro Spallanzani showed that chemicals made in the stomach were important in breaking down food into a simpler form. He took liquid called gastric juice from the stomach and mixed it with food in a jar. The food was broken down in a similar way to digestion.

The main problem that the early scientists had was how to find a way to look at the process of digestion as it happened inside a living body. They could not think of an experiment that they could do on a living person to investigate the process.

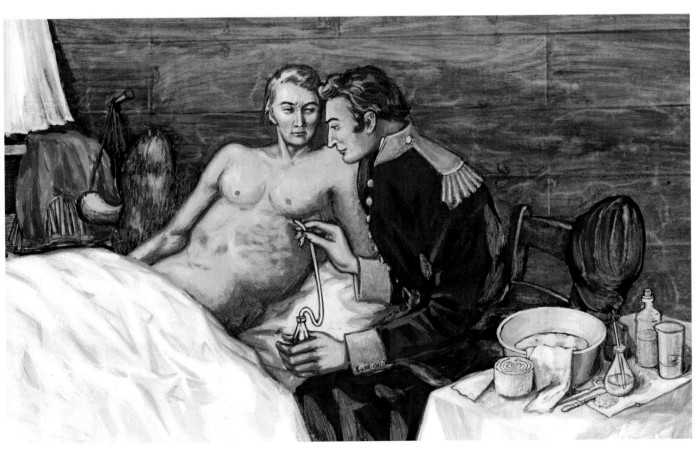

Dr William Beaumont and Alexis St Martin.

Enter Dr William Beaumont

In 1822, William Beaumont was a United States army surgeon working at a camp in Michigan. On the 6th of June he was called to treat a young man, Alexis St. Martin, who had been accidentally shot at close range in the chest. Alexis had been working at the camp, where he collected furs from the Native Americans and delivered them to the local trading company.

The wound, according to Dr Beaumont, 'was caused by a charge consisting of powder and duck shot that was fired from close range into the left hand side of the trader'. It left a large hole in his side. Out of this hole protruded a burnt portion of the lung, as large as a fist. Below this something else was sticking out. Closer inspection showed that this was part of Alexis' stomach. The stomach had a hole in it as large as a finger, through which poured the food the young man had eaten for breakfast!

Dr Beaumont put back the stomach and lung as best he could, and cleaned and dressed the wound. He did not expect Alexis to live, but intended to make his end as comfortable as possible.

To everybody's surprise Alexis not only survived but he made a most remarkable recovery, almost to normal health. He could not keep food down, however, because of the gaping wound in his side. Dr Beaumont continued to treat the wound, but despite his best efforts, it would not heal completely. Alexis was left with a hole in his side that led into his stomach!

Gradually a lid of skin grew over the hole. This lid could be opened, allowing the inside of Alexis' stomach to be seen. Alexis could live quite normally with this wound, but could not continue his work as a fur trader. Dr Beaumont gave him work as a handyman in his home.

Table of results from Dr Beaumont's note book.

Article of Diet	Mode of Preparation	Time required for digestion (Hours/Mins)
Eggs	Raw	2.00
Eggs	Soft Boiled	3.00
Eggs	Hard Boiled	3.30
Trout	Boiled	1.30
Cod	Boiled	2.00
Salmon,	Boiled	4.00
Salted Potato	Boiled	3.30
Potato	Baked	2.30
Pork,	Raw	3.00
lean Pork	Fried	4.00
lean Beef	Fried	4.00
Beef	Boiled	2.45

The human guinea pig

In fact, Dr Beaumont was interested in the different ideas about what happened to food in the stomach before he met Alexis St Martin and he had read about the ideas of other scientists. Being curious, as all scientists are, he now saw an opportunity to investigate what was going on in Alexis' stomach. He designed a series of experiments that could help solve the riddle of digestion.

Dr Beaumont began investigating the inside of the stomach and its secretions. He tied a piece of silk around samples of food and put these through the hole into the stomach. He wanted to know what happened to food inside the stomach. He tried different types of food, including raw, sliced cabbage, a piece of stale bread, raw pork, cooked beef, raw salted lean beef, boiled salted beef, and many others.

At hourly intervals he removed each food sample and examined it to see what had happened to it. After the first hour he found that the raw cabbage and stale bread had been half digested, while the pieces of meat remained unchanged. After two hours the cabbage, bread, pork, and boiled beef were all completely digested, but the other meat was very little affected.

Dr Beaumont performed hundreds of similar experiments on Alexis over the next eight years. Some of his results are shown in the table on page 10.

In another experiment, Beaumont extracted some of the gastric juice from Alexis' stomach. He put food samples into a jar of this juice. He compared how quickly the food was digested in the jar with the time it took for food to be digested in the stomach. He tried this at different temperatures and found the food in the jar was digested at about 37°C.

Dr Beaumont made two very important discoveries. He showed that gastric juice, when placed in a glass jar, would dissolve food at just the same rate and in just the same way as they were dissolved inside Alexis' stomach. He also showed that the important acid in gastric juice was hydrochloric acid.

The puzzle that had baffled scientists for hundreds of years was finally solved.

William Beaumont continued his studies into digestion, but chemical analysis at that time was limited and it wasn't until many years later that scientists discovered that the important chemicals in the juices made by the digestive system were enzymes.

As for Alexis St Martin – he married and had several children, and lived on to a healthy old age, and was perhaps known to his friends as – 'the man with the hole in his stomach'.

QUESTIONS

1. Read carefully how Dr Beaumont designed his experiments into digestion.
a) What steps would he have to take to make sure that he carried out a fair test when comparing the different foods?
b) What might he have done to make sure he did not infect the wound?
2. Look at the table showing some of Dr Beaumont's results.
a) Draw a bar chart to show the results.
b) Describe any patterns that you can see in the results.
3. Imagine that you are Dr Beaumont. Write a letter to your wife explaining how you felt when you realised how Alexis St Martin's injury could help solve a scientific problem.
4. Imagine that you are Alexis St Martin. Write about what it felt like to be used as a human 'guinea pig' in Dr Beaumont's experiments. Perhaps you are proud to take part, or maybe you feel that you have no real choice since you can no longer work for a living.
5. Describe how Dr Beaumont's experiments changed scientists' views about how food is digested.

The body's chemical factory

An adult's liver is about 15 cm thick and weighs 1.5 kg. It's like a chemical factory, removing worn-out cells, neutralising harmful substances, producing and releasing useful substances for the body. Its job is to keep the internal environment of the body as constantly balanced as possible.

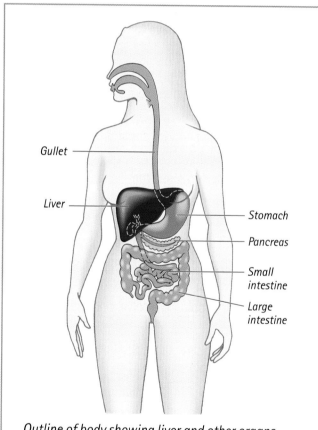

Outline of body showing liver and other organs of the digestive system.

Gullet
Liver
Stomach
Pancreas
Small intestine
Large intestine

Hi, I'm Libby the liver cell. Your body is organised like a small town. It is made up of small units that use materials and produce waste.

These waste materials are found in the blood and have to be got rid of or, in science terms, excrete If the waste builds up it will eventuall kill us and you!

1

Each small part has an important role, which helps the whole thing work. You eat food. Friends of mine in the stomach break your food down into more useful bits. Other cell friends of mine transfer energy from the chemicals in the food, like carbohydrates and fats. This produces waste.

You can find me in one of the most important and largest organs in the body – the liver. Of course, people have known for a long time that the liver is an important organ.

2

The Greek doctor Hippocrates thought blood originated in the liver. He also thought it gives people violent feelings or emotions. Others believed the liver was the place in the body where love originated! The word liver comes from an Old English word 'lifer', meaning 'life'.

Our most important job is to break down and recycle old blood cells. During the breakdown we produce chemicals that are returned to the blood or stored in the liver as a mixture of substances called bile.

Bile also includes chemicals we make to help digest fats in the small intestine – and which gives our solid waste its brown colour.

3

It really is an unpleasant mixture, so we send it to the gall bladder (a small sac next to the liver) for storage.

Sometimes, crystals form in the gall bladder. These are called gall stones and they can block the gall bladder, so that the bile liquid leaks into the bloodstream. The bile gives the skin and eyes a yellow appearance. This condition is called jaundice. Luckily, the stones can be removed quite easily.

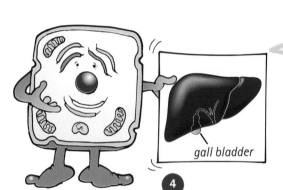

gall bladder

4

We liver cells act like a bank, collecting and storing vitamins from your food. You remember being told to eat carrots because they are good for you? Well, the liver uses the orange pigment carotene to make vitamin A, necessary to help you see in dim light.

We also make and store a high-energy chemical called glycogen, which can be converted to a sugar called glucose. You sometimes need to do this when the glucose available to your muscles runs too low, such as in a marathon race or long periods of strenuous activity.

5

We sometimes cause problems for athletes when we break down certain chemicals they take to help cure common illnesses. This is because we can turn them into other chemicals that look like banned drugs

One of the more controversial jobs we do in the liver is to change fatty substances so that they can travel easily in the blood. You may have heard of cholesterol as a substance you should eat in small amounts. Well, we try to control the amount of cholesterol in your blood by either taking it out of your bloodstream or pushing more back in!

One of the most important of our 500 or so jobs is to remove poisons or toxic substances like alcohol. But we can be damaged by large amounts of alcohol and the painkiller paracetamol. This damage can be very bad and lead to our death or make us behave abnormally.

6

We cover ourselves with a sort of skin that stops the liver from removing toxins and doing its normal job. The liver becomes very big and the person can get the disease called cirrhosis.

To cure this disease, parts of the liver must be removed. In this operation the adult liver can be cut down to one-tenth of its usual size and still work normally.

7

The size of the liver is related to the size of the person and how energetic they are . So a large and active person will have a large liver and a small inactive person will have a small liver.

If the liver is cut down in size the slow replacement of liver cells suddenly speeds up. This fast growth of liver cells also happens when a woman becomes pregnant.

8

Surgeons make use of the liver's fast regeneration when they transplant a small part of an adult liver into a younger person. The adult's liver regenerates very fast so they are not left feeling ill. The young person's liver regenerates, so giving them a working liver that lets them live a normal life. Interestingly, surgeons have noticed that large livers also shrink to suit the size and activity of the person. One famous scientist who studied the liver was the Frenchman Claude Bernard (1813-78). Claude Bernard was not seen as a good student. In fact he found his studies so boring he decided to write plays instead. However, his plays were not successful and he turned to studying the body and its processes – and he became an excellent scientist.

Claude Bernard.

Using dogs, Bernard discovered that carbohydrates are broken down into sugars during digestion. He also found that the liver can make a carbohydrate and that fats are broken down by bile juice. He worked like a modern scientist, making use of careful observations and looking for patterns in the results without trying to fit them to an idea. His most famous idea is that life relies upon cells working together in a constant environment. These cells work in a narrow range of differing conditions such as temperature or concentration of chemicals. This led to the idea that organisms can control their internal environment – just as we have been describing here!

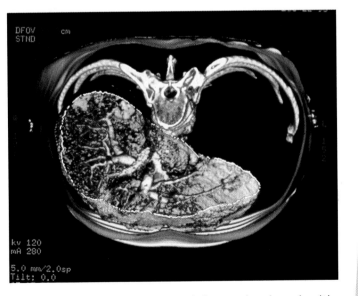

Cross-section through a human abdomen showing a healthy liver (pink) below the backbone and rib cage (blue). Red arteries run across the surface of the liver. To the left of the backbone is the gall bladder (green).

QUESTIONS

1. Draw a diagram of the body and show on it the position and size of the stomach, kidneys, lungs, heart and liver.
2. Find out more about Claude Bernard and Ernest Starling. Where were they born, where did they live and why is their work on the human body connected? Write a biography of them for a science magazine, discussing the connection between them.
3. Some people are organ donors and carry donor cards. We have seen that organ donation does not always mean being dead to donate an organ. Write a newspaper article giving the good reasons for organ donation.

Arguments about microbes

This is a story about a public quarrel and a public experiment. For hundreds of years people had believed that the maggots and insects you could find in bad food just happened. They did not come from somewhere else, they were 'generated'. Once the microscope was invented it was possible to see even smaller things moving about in bad food and bad water. These were called microbes or micro-organisms. They were obviously alive, but where did they come from?

This was a very important question. Bad food is pretty smelly and disgusting, but it can also spread disease because it contains micro-organisms. These live on the bad food and make poisonous chemicals – toxins – from it. If micro-organisms get into our bodies they can multiply very fast until they are making so much of these poisonous toxins that we become very ill. Just eating the toxin can be enough to poison us. Even today many thousands of people get food poisoning from bad food every year, and some of them even die. Are these dangerous microbes just created out of nothing, inside our food?

In 1859, a young French scientist called Louis Pasteur (does that name remind you of milk?) decided that he would do a careful experiment to show that the spontaneous generation of microbes just could not happen. And he wanted all the world to hear about it – he was that sort of person!

Pasteur's hypothesis was that spores (rather like eggs) from which microbes could grow just floated about in the air. They were on every surface and in every piece of food, unless it was sterilised by boiling. And yet they could not be seen except under the microscope. One useful thing was that when clear meat soup went bad you could see that, because it went cloudy. However, just sterilising soup and showing that it stayed clear until more air was added was

Louis Pasteur, a chemist and the founder of microbiology, works on an experiment.

not enough to convince some of the other scientists. They just answered that microbes needed fresh air to grow properly, but they did grow out of nothing. So now Pasteur made his plan for a very careful series of experiments with different kinds of air.

Pasteur began by making some special apparatus. It was a glass flask with a long neck and some liquid inside. Pasteur then heated the neck of each flask and pulled it out into a long S-shape. When the soup was boiled steam came out, sterilising the 'swan-neck' of the flask. Then he let the soup cool down. Air came in but the microbes were trapped in the bend of the tube. His soup did not go bad. Then he melted the end of the tube and sealed it up.

Now came the publicity act! Pasteur prepared 20 of these swan-necked flasks, all sterilised and sealed. He got together mules to carry them and assistants to help him, and set out for the mountains. Everyone knew about his experiment!

At the bottom of the mountains, where there were farms and animals, Pasteur stopped his line of mules, took down six of the flasks, broke open the end of the tubes so that the air could rush in, waited 20 minutes, and then sealed the flasks. Three days later the soup in all six flasks had gone bad.

Pasteur prepared a flask of soup and heated the glass neck of the flask so that he could pull it out into an S-shape. Pasteur boiled the soup and left the end of the neck open to the air. Would the soup go bad?

Then the whole troupe of mules and equipment set off again up the mountain. About halfway to the top Pasteur opened another six flasks, let the air rush in, waited 20 minutes, and then sealed them up again. In all except one of the flasks the soup went bad.

Then he went on up to the snow fields and glaciers on the top of the mountain. Here there were no animals and few people. Pasteur opened six more of the flasks and the air rushed in. Once again he sealed the flasks and waited. This time the soup only went cloudy in one of the flasks!

'There you are,' thought Pasteur, *'when there is pure air, with no microbes in it, the soup does not go bad. That means microbes are not just generated inside the soup. They do come from the air and the cleaner the air the less the numbers of microbes in it!'*

But some people are very hard to convince. Have you noticed that even when you have a really good argument they refuse to change their minds? Scientists are no different at the beginning. But experiments do, in the end, provide good evidence and change scientists' opinions. Pasteur asked the French Academie des Sciences to do the same experiments that he had done. They agreed, and no doubt Pasteur had a few sleepless nights waiting for the results. But when it came, the news was good. The same thing had happened again. All the scientists in the Academie agreed that microbes came from the air. Spontaneous generation was impossible!

Very soon all the scientists in Europe agreed with Pasteur. His experiment had been so convincing.

QUESTIONS

1. Have you, or anyone in your family, ever heard of a case where modern scientists have disagreed with each other? If so, describe briefly what the disagreement was about.
2. Which do you think are better – arguments heard in public or secret arguments between scientists only? Give your reasons.
3. Why is it important that all experiments should be repeated?
4. Why do you think that in one of Pasteur's flasks the soup went bad even at the very top of the mountain? Try to give more than one reason for this, and say which one you think is more likely to have been right.
5. Look up 'pasteurised' and explain how it is connected with Pasteur's work.

Jabs for James

In the eighteenth century smallpox was a killer disease. There was no cure for anyone who caught it. Then a doctor called Edward Jenner noticed something interesting. Milkmaids, who often caught a mild disease called cowpox from the cows, did not get smallpox.

Dr Jenner had the idea that the cowpox somehow 'got in the way' of smallpox. He wanted to do an experiment to test out his idea. For his experiment he chose a small boy – James Phipps – who had never caught smallpox. First, Dr Jenner got some pus from a spot on the hand of a milkmaid, called Sarah Nelmes, who had cowpox. He then made a small cut in James's arm and rubbed in the pus.

About five days later James became ill with cowpox, but he soon recovered.

Dr Jenner then took some pus from the spots on a smallpox victim. Again he made a small cut in James's arm and rubbed in the pus of this deadly disease. Dr Jenner waited, but nothing happened. James did not get ill. Dr Jenner had successfully protected him against smallpox!

Dr Jenner called his method of protecting against smallpox 'vaccination' after the medical name for cowpox. Today, this method has been so successful that doctors have managed to wipe out smallpox throughout the world.

The smallpox debate

The extracts below describe some of the characters closely involved in the story of Jenner's vaccination experiment. Read through each extract.

(a) If you are working in a group you may like to take the part of a particular character and then be prepared to 'role-play' that character in a press conference. Some pupils can take the part of a group of journalists and ask you questions about your attitudes to the experiment.

(b) If you are working on your own, then use the 'Now decide' questions to describe how you think each of the characters would have felt about the experiment.

JAMES

You are eight and three-quarters years old and love fishing. You have gone to Sunday school a few times but you usually manage to escape and go down to the river. You don't want to learn to read. You don't want to start work either, but you know you will have to as soon as you are nine.

When you were little you fell out of a tree and broke your leg. Your father got Dr Jenner to set the broken bone and it hurt very badly. You have always been scared of

Dr Jenner from that time. Once, he gave you a medicine made from bitter aloes to cure your fever. You couldn't get the taste out of your mouth for a week afterwards! You don't know much about smallpox except that your aunt had it. Now her face looks horrible with large deep pits all over it. You can't even bear to kiss her.

NOW DECIDE

1. Did you understand what Dr Jenner was going to do?
2. How would you feel about asking the doctor questions?
3. Now that you are safe from having smallpox would you advise your friends to have the vaccination too?

JAMES'S MUM

You did not go to school but have worked with your husband in the fields for many years. Now that he is gardener to Dr Jenner things are much easier.

You've had nine children but two of them died when they were only babies. You have never forgotten that. Dr Jenner did come and give the babies some medicine, but it did not help them. He was not able to help your sister either when she died in childbirth.

James is your youngest and you know you spoil him a little, but you can't help it. He's always out fishing when your husband wants him to help in the garden. A month ago your husband suggested that James should be given smallpox now to prevent him from getting it later. That seemed terribly dangerous. You didn't sleep for a whole week worrying about it.

NOW DECIDE

1. How do you feel about doctors?
2. What did you think when Dr Jenner explained what he was going to do to James?
3. Did you talk it over with your husband? If so what did you say to him?

JAMES'S DAD

You never went to school and have worked on a farm since you were very little. It has been very tiring work. Now you are nearly fifty, as near as you can tell. You have an easier job as Dr Jenner's gardener and you are very pleased. He is a very good employer.

You grow vegetables for Dr Jenner, and he allows you to take some home. He also lets you and your family live in the cottage by his gate. You have had nine children, but two of them died as babies. That's quite common in the village.

James is your youngest son and he is nearly old enough to go out to work and bring in some more money for the family. James has always been healthy, but you think he is lazy. You have discussed whether he should be given the smallpox now, so that he will not get it later. Your wife was against it.

NOW DECIDE

1. What are your feelings towards Dr Jenner?
2. What did he say to you about vaccinating James?
3. What did you reply, and what did you secretly think about it? How did your wife react?

SARAH NELMES

You are 13 years old and have been a milkmaid since you were eight. It is long, hard work. Your family are poor and have always lived in the village

Just a week ago you caught the cowpox. Most of the other milkmaids you know have had it too. It hurts and you have seven large blisters on your hands, which have gone yellow with pus. You didn't worry too much about it because your friends tell you it will go away in about a week. No one ever dies of cowpox.

You know Dr Jenner. He looked after your mother when she was very ill last year, without asking for any payment. You think he is a kind and very clever man. He can read and write and has even been to London.

NOW DECIDE

1. What did you think when the doctor said he was going to cut into one of your blisters?

2. Do you trust Dr Jenner?

3. What questions did you ask him?

DR EDWARD JENNER

You were born into quite a wealthy family. When you were only 13 you began work apprenticed to a doctor, and you are now 47 years old. When you were very small your parents arranged to have you given smallpox, but you only had it mildly. Now you are safe from smallpox. You are 'immune' to it.

You have been interested in smallpox for at least seven years and have been making notes about all the cases you could find of people who had cowpox and then did not get smallpox. You are convinced that if you have cowpox you cannot get smallpox.

You did try an experiment six years ago. You put pus from a smallpox victim into the arms of several people who said they had already had cowpox. Some of them became very ill with smallpox.

These results puzzle you. Even though other scientists laugh at your ideas about vaccination, saying people might turn into cows, you want to try again.

NOW DECIDE

1. Did you do this experiment to get scientific fame or to help people?

2. How did you persuade Sarah to let you take pus from her arm?

3. How did you persuade James's parents to let you use James for your experiment?

4. When James was ill, were you worried about what might happen?

Jenner the scientist

The words below describe some of the processes that are involved in developing scientific ideas.

Process	Meaning
OBSERVING	To watch carefully what is happening. (As well as using your eyes you can observe by listening, smelling, touching and occasionally tasting.)
MAKING A HYPOTHESIS	To have an idea about why something happens. (You can use this idea to design experiments.)
DOING AN EXPERIMENT	When you design and carry out a test of your hypothesis.
PREDICTING	Saying what you think is going to happen during your experiment.
REACHING A CONCLUSION	Deciding what your experiment shows. (You should be able to tell if your hypothesis is right or wrong.)

Now copy out the table below and decide which scientific process Dr Jenner was using.
Use the process words from the table above.

What Jenner did	Process
Jenner puts cowpox pus into James's arm	
Jenner sees that James doesn't get smallpox	
Jenner hears that the milkmaids don't get smallpox	
Jenner decides that having cowpox stops you getting smallpox	
Jenner sees that James suffers from cowpox for a few days, and then gets well	
Jenner takes some pus from a smallpox victim and puts it into James's arm	
Jenner thinks that if he gives someone cowpox first they won't get smallpox	
Jenner hears that milkmaids often get cowpox	
Jenner takes cowpox pus from a milkmaid	
Jenner decides that having cowpox will stop you getting smallpox	

Mary Mallon – typhoid carrier

Typhoid fever is a serious illness caused by a bacterium, which results in diarrhoea and vomiting. It is common in areas where raw sewage can mix with water meant for drinking or cleaning. It causes many deaths in places without proper sewage treatment. It is a common cause of death after natural disasters such as earthquakes.

At the start of the twentieth century, typhoid was a major cause of death in even the most advanced countries. More than 35 000 Americans were killed by it in 1900.

One outbreak of typhoid occurred on August 26th 1906, in a house in Oyster Bay. Charles Warren had rented the house for a family holiday. Six of his family suddenly became ill. The public health officials moved in at once to investigate. They tested the water supply for contamination with typhoid bacteria. They were baffled when they found none, and wondered what could be the source of the disease.

The owner of the house was determined to find the source of the infection. He was worried that he would not be able to rent it out again the following summer. He called in Dr George Soper, an expert on typhoid, who had a reputation for detective work.

Soper was a scientist. He worked in a methodical way. He inspected the drains. He tested the milk the family drank. He tested the clams they ate. All of these tests got him nowhere. He couldn't find any typhoid bacteria.

Next, Soper tested the servants and found his first clue. The Warrens had hired a new cook on August 4th, just three weeks before the illness struck. Three weeks is the incubation period for the disease.

He called for the cook but she had disappeared soon after the disease broke out. He became even more suspicious. All he could find out about her was her name – Mary Mallon.

Determined to solve the mystery, Dr Soper went to the agency that had hired Mary for the Warren household. He made a list of her previous employers, and noticed a pattern.

Almost all of them had suffered outbreaks of typhoid soon after employing a new cook. In each case the cook moved on soon after the outbreak. In each outbreak, the cook did not become ill. Soper could explain this.

He had been studying the latest scientific research into the disease. New evidence from scientists in Europe showed that it was possible for some people to carry a disease without becoming ill themselves. A carrier is immune to the germs, but they can breed on the person's body and be passed on by them.

Dr Soper's notes

New York, 1900.
Mary had worked as a family cook.
One of the family became ill with typhoid.

New York, 1901.
Mary had moved to another household. Shortly after, one of the kitchen staff died of typhoid.

Dark Harbour village, 1902.
Mary again worked as a cook for a family. Eleven of the family became ill with typhoid.

Between 1903 and 1906.
Mary worked as a cook for four different families. Each family suffered outbreaks of typhoid.

Soper was convinced that Mary was a carrier of typhoid bacteria. He was convinced that she did not wash her hands properly after using the toilet. Her hands became covered with bacteria, which were passed on to the food she prepared. Innocent people eating the food would become ill after being infected.

*I*t was vital for Soper to find Mary. She could cause the deaths of many more people. In 1907 he tracked her down to her latest job. She was working as a cook again, this time for a family in Park Avenue, New York.

*D*r Soper told the head of the family about his suspicions. Mary was called for. He told her that she was carrying typhoid, and asked her to allow him to carry out tests. She was furious. She did not believe his theory at all.

*S*he refused to co-operate. She picked up a fork and chased him from the house.

*S*oper told Dr Josephine Baker at the Board of Health. She called on Mary but she was also chased away. The next day, she returned with the police. Mary became frightened and fled.

*T*he police soon found Mary hiding next door. They took her, fighting and kicking, to hospital in an ambulance. Dr Baker had to sit on her to keep her down!

*L*aboratory tests proved Dr Soper's theory was correct. Mary was a carrier of typhoid. Her body was full of the bacteria. The doctors did all they could, but she refused treatment. She could not believe what they were telling her.

The police would not let her leave the hospital. She was sent to live in an isolated house in its grounds. Everything was done to make her comfortable, but she was, in effect, a prisoner. She was very unhappy with her fate. She felt it was just not fair. In 1910, the police agreed to release Mary from the hospital. She promised never to work in kitchens again.

Mary Mallon was not heard of for five years. But there were more outbreaks of typhoid. Some were traced to kitchens. In no cases was the cook called Mary Mallon.

In 1915 typhoid broke out in a New York maternity hospital. Twenty patients became ill with the disease. Doctor Soper was again called in. He sent for the hospital cook – it was Mary Mallon! Mary ran away again. The trail was hot this time however. She was soon captured by police in Long Island, New York. They were shocked to find her serving food even as they burst in on her!

QUESTIONS

1. Why did Mary not suffer from typhoid herself?
2. Explain what a 'carrier' is.
3. Make a list of the pieces of evidence that suggested that Mary was a carrier of typhoid. How reliable do you think each piece of evidence is?
4. Why do you think Mary did not believe Dr Soper's theory?
5. Why was it necessary to keep Mary a 'prisoner' in hospital? Do you think this was fair treatment? What would happen to similar cases today?

For the next 23 years Mary was kept imprisoned in hospital. She was given work in the hospital but she was never allowed to prepare food. She remained angry and refused all friendship, eating alone. She died in 1938, still a carrier of typhoid bacteria.

Peanut allergy

Some people, adults as well as children, have an allergy to peanuts. This means that their bodies just cannot cope with a particular substance in the peanut, even in tiny amounts. On one occasion, a little boy aged less than two was kissed by an aunt who had just eaten a peanut sandwich. Just that kiss was enough! The boy went into shock. It was as though he had eaten a poison. He began to breathe in a very shallow way, his heart beat slower, his skin felt colder and all his muscles became slack. He very nearly died from that kiss! The only way the doctor could make him better was to give him an injection of adrenaline.

Adrenaline is a substance that we all produce when we are excited or frightened. It makes all parts of our body work faster. We breathe quicker, our hearts beat more strongly, our muscles become tense and we may be able to run away from danger faster than usual. You can see that it is just the right substance to help people who are in shock. The only problem is that those with a bad peanut allergy have to carry around a syringe full of adrenaline wherever they go. The injection of adrenaline cures the shock, but the patient feels very tired and sick afterwards.

Hope of a cure?

Scientists have made some progress. Firstly, they bred a special kind of laboratory mice that are also allergic to peanuts. Just like humans they go into shock if they eat even the smallest bit of a peanut. Secondly, the scientists have found the gene in peanuts responsible for making the deadly substance, called an allergen, that affects people who have the allergy.

Then the scientists did a very odd thing. They took the peanut gene and fed it to the special mice so that they would start making the deadly allergen. Sure enough, some of the cells in the digestive systems of the mice began to make small quantities of the chemical, just as if the mice were peanuts! But the mice seemed quite well and happy. They did not go into shock. Was the substance – usually deadly – slightly different because it had been made inside their own bodies?

You can sometimes inject yourself with adrenaline to avoid the effects of eating nut products when you have an allergy to nuts.

Surviving the peanut test

Now the scientists had two sets of mice. They looked just the same. Both originally had an allergy to peanuts, but the second set had made small quantities of their own peanut allergen without getting ill. Had they been immunised against the allergy?

The scientists decided to carry out a controlled experiment. Each lot of mice were fed peanuts. The first lot went into shock as usual; they became very ill, only breathed from time to time, and did not respond if they were gently prodded. Some died. But the second lot of mice had very little reaction at all to the peanuts. It was as if they had been vaccinated against the allergy by the allergen they had made, just as we can be vaccinated against mumps and measles.

The scientists are hoping that, one day, they may be able to vaccinate children who have the peanut allergy to stop them going into dangerous shock. However, they say that there are other experiments that they must do first.

Mouse with peanut allergy   Mouse with peanut allergy

Feed mouse with gene for peanut allergy

Feed peanuts to mouse (mice)

Mouse dies   Mouse healthy

Mice can be vaccinated against the peanut allergen.

QUESTIONS

1. What evidence is there that the laboratory mice have the same peanut allergy as humans?
2. Suggest two ways in which adrenaline produces the opposite effect to the peanut allergy.
3. Describe how you think the scientists carried out the 'controlled experiment' to find out if the two sets of mice behaved in different ways when they ate peanuts.
4. Do you think the scientists were right to use live mice in these experiments? Write down your own ideas before discussing them as a class.
5. Suggest one experiment you think the scientists might want to do before they make a vaccine for children.

Poisons and medicines from plants

'Wise women', whom some people called witches, used to collect wild plants to make medicines. Even doctors used parts of plants to treat their patients. Foxglove leaves were used for heart trouble, and we still use the substance called digitalis, extracted from these leaves, today. Some plants had leaves that were used as a compress on the skin to take away the pain, rather like the dock leaves we still use to stop the pain from a stinging nettle. You can be sure that every wild flower with a name ending in 'wort' (like woundwort and St John's wort) was once used to make people better.

Wise woman or witch?

Both doctors and 'wise women' knew about the poisons they could get from other plants, like deadly nightshade. They knew much more about herbs than most people did.

St John's Wort. The leaf tips and flowers are used to treat a variety of ailments, including depression.

Woundwort.

Sometimes a person would come to buy one of these poisons in order to make a drink to harm or kill someone they didn't like. So if an ill person went to a 'wise woman' for a cure, they might be frightened about what they were getting. If they died, their relatives might believe that they had been given a poison, or that the woman had used magic against them. Then sometimes they came to burn the woman as an 'evil witch'. In those days it was very dangerous to be an old woman living alone in the woods with only a black cat for company!

Foxglove.

Amazonian Indians tip their arrows with curare to paralyse the animals they hunt.

In South America a special plant poison called curare was used in hunting by the Indians of the Amazon. It came from a vine and looked like a sticky brown mess. The Indians dipped the tips of their arrows into this mess before they went out hunting. Then the arrow had only to graze the skin of an animal and the poison started moving round inside its body. One after another, the muscles of the body relaxed until the animal fell over and could not move. The poison first affected the toes and eye-lids. Then the neck and legs went limp and could not move. After that, the curare reached the muscles of the diaphragm, which lifts the lungs to push out the air and then pulls back to let it in (can you feel yours just below your ribs?). When the poison arrived here the diaphragm stopped working. The animal could no longer breathe and death soon followed!

If a person or animal swallowed curare it did them no harm. To paralyse them it had to get into the bloodstream. This was useful because it meant that the Indians could cook and eat meat from the animal they had killed without being poisoned themselves.

Claude Bernard's frogs

When curare and other plant poisons were brought to the University of Paris in the nineteenth century they were investigated by Claude Bernard. He was very interested in all poisons and had a hypothesis that every poison might be different and might act on different organs. He did not believe that each one just knocked out the whole body. He began by carrying out some experiments on a frog, the body of which had been paralysed by a mild dose of curare. All the muscles seemed to have stopped working. Had they been directly damaged by the curare, or was it something else?

To find out, Bernard gave the frog's leg a gentle electric shock. The muscles contracted and the leg kicked! This meant that the muscles were alright even though the leg was still paralysed. What was keeping it like that? Why were the messages that normally travel from the brain through the nerves to the muscles not making the leg move?

In his next experiment he tied a string round one of the frog's legs before giving it a mild dose of curare. This time

29

the lower leg wasn't affected – the frog could still move it. Claude Bernard guessed that the poison had not reached it. By careful measurements he found that it was only the junction between the nerve and the muscle that was affected. There was some special tissue there and the curare had stopped it from passing on messages from the nerves. So the experiment supported Bernard's hypothesis – there was only one place where the poison was doing damage. After a while, the frogs recovered fully and all their muscles worked again.

Curare and surgery

Could curare be of any use in medicine? It did occur to some doctors that stopping the muscles from contracting would be valuable during operations, but natural curare was still a dangerous poison. Some samples of it were stronger than others. That's the trouble with plant medicines, you can never be sure how strong they will be. It all seems to depend on the particular plant and the amount of rainfall there has been recently. Then a chemist had a brilliant idea. Why not combine the curare with acid to make a new substance, rather like a salt, which could be purified? Then doctors would know exactly how strong a few grams of the drug was.

The new drug they made in this way was called tubocurarine, but it was not made pure enough for reliable use until 1935. Since then it has been used in anaesthetics to make patients' muscles relax. This only lasts for about 20 minutes, after which more anaesthetic may be required. Of course, while a patient is so relaxed with tubocurarine they cannot breathe by themselves, so artificial respiration is required. A tube is put into the patient's mouth and air is pushed into their lungs at regular intervals. (You have probably seen this in hospital dramas on TV.) Meanwhile, the patient is quite unconscious and their body is totally relaxed.

There are other special uses for curare. One of these is to help people whose muscles keep jerking painfully. Another is to relax the patient's throat so that a miniature camera in a long tube can be lowered down to see if there is an obstruction in it, like a fish bone or a growth, which needs to be removed.

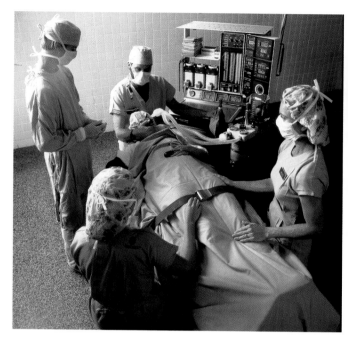

Anaesthesia being administered prior to surgery in a US hospital operating theatre.

This anaesthetic does have some disadvantages. The patient's muscles feel rather weak for several hours or even a day after the operation, and also the bronchial tubes that branch into the lungs may become constricted for a little while. But now there are other substances that can be added to make the curare work better. So it's good to know that the plant poison that was once used in the forests of the Amazon to hunt animals is still helping surgeons all around the world perform operations safely.

QUESTIONS

1. Describe a medicine you have had or heard about that was made from herbs (natural ingredients).
2. Why do you think people believed that the 'wise women' who gave them herbal medicines might be 'evil witches'?
3. When a patient is being given tubocurarine before an operation, will they be given it in a pill or as an injection?
4. Why do patients having this anaesthetic also need artificial respiration during their operation?

Frankenfoods or saviours of the world?

We are on the threshold of a great revolution. The world as we see it now will be unrecognisable in 20 years time. This is more dramatic than the invention of the computer, or electricity, or the steam engine or the plough. This is the dawning of the 'gene revolution'. Are you ready for it? Do you know whether it's a good thing or a bad thing? Do you think there ought to be limits on genetic research or do you think all will come right in the end?

Why not hold your own class debate?

- Choose equal numbers of speakers on both sides of the debate. They should research the topic using the information on these pages or the Internet ...
- When you're ready, the main speakers should line up in front of the class, with a chairperson to ensure no rowdiness or mega-long speeches. (It's best to put a time limit on it – tell the speakers they only have two minutes, and give them a warning when they've only got 30 seconds left.)

> **The Motion Is:**
> 'We believe the gene revolution will bring benefits to everyone.'

- The main speakers alternate, for and against the motion. The chairperson should ensure that each speaker gets a full, uninterrupted two minutes to put their case. Then the debate is open to 'the floor' – others in the class can add their comments.
- Take a vote among the audience before and after the debate, to see if the speakers have changed any minds.

On the next two pages are some of the arguments for and against the so-called 'gene revolution'.

FOR

1 About 10 years ago scientists worked out how to 'read' the genetic code, so now they can read the basic blueprint of any living thing. Now we have a complete map of the human blueprint (the 'human genome') we will be able to check parents and babies and help deal with life-threatening illnesses before the baby is even born. This is a life-saving technology.

2 By manipulating the genes of a plant or animal, adding new ones, or taking others away, modified crops can be produced. For instance, people in China who eat little more than rice suffer from a lack of iron. By adding an extra gene, the rice can produce more iron, curing the farmers' problems.

3 There are a lot of scare stories about 'Frankenfoods', but scientists are too sensible to allow mad experiments to be carried out. Nobody will be allowed to clone humans, for instance.

4 The gene revolution is impossible to stop. What we must do is be careful that mistakes are not made: there need to be controls and proper tests to make sure that there are no harmful side effects. Provided that happens, nothing can go wrong.

5 Many illnesses involve parts of the body that are faulty. For instance, severe burns destroy skin. But you can't easily replace it with someone else's skin – the body rejects the graft. With a little genetic manipulation, a fresh piece of your own skin can be grown in a laboratory, making the healing process so much easier. The same can be done for other body parts – bone, muscle or cartilage, ears, or even livers and kidneys.

6 The basic idea behind gene manipulation has been around for thousands of years. Way back in 7000BC, farmers saw that some cows gave more milk than others. They were keen to breed from these cows to increase milk yields. Over thousands of years this produced the modern dairy cow, with udders so large that the animal could no longer survive in the wild, but depends on humans. Likewise, wheat was developed from wild grass, modern sheep came from wild sheep, etc. Genetic modification does nothing more than make this process faster and more efficient.

7 World starvation could be only just around the corner. Food production increases every year at a steady rate, but human populations increase much faster. Without some sort of help, the food supply won't be able to keep up.

8 There have been many agricultural revolutions in the past: the domestication of crops about 9000 years ago, the invention of the plough 5000 years ago, the agricultural revolution of 200 years ago, the agrochemical revolution fifty years ago, all boosted food production. Genetically modified crops are the latest revolution.

9 The use of built-in resistance to pests should mean a reduction in the use of harmful sprays, and a cleaning up of the countryside.

10 It will be possible to clear up industrial sites polluted with highly toxic chemicals using genetically modified bacteria that consume the chemicals, then render them harmless.

11 Throughout history every new development has been accompanied by screams of fear from those who are scared of it. It is natural to be wary of new developments, but in the end we must realise that more good than harm will come from GM foods.

12 Attacking GM research could have serious side-effects. Fears over some more controversial tests has led to the collapse of traditional ones, e.g. trees are being modified to produce better wood-pulp for papermaking. This will result in fewer forest trees being cut down. But protesters destroyed the experimental trees before the trial was completed, so the modified trees may never be produced. Vaccines to protect people against certain diseases could be grown in genetically engineered bananas or potatoes. Tests for these can be conducted safely in greenhouses, yet investors are reacting to the public outcry by withdrawing funds, putting the tests in danger.

AGAINST

1. Many new inventions are not as useful as they are made out to be. Scientists will always exaggerate the benefits of their latest ideas, and play down the drawbacks. Nobody can foresee the future side effects of GM foods. A long test period is essential for such a potentially dangerous process.

2. With genetic manipulation we are stepping into an unknown area. If a new form of life is created, what effect could it have on the rest of nature?

3. Of course new GM products are tested for safety, but how can we be sure that they are properly done? Testing products is expensive! It could be very tempting to cut a few corners and 'fiddle' the results in order to get the product on the market.

4. Some of the companies involved in GM foods are enormously powerful and are making higher profits than the income of some countries of the world. It could be that they will put undue pressure on governments to allow their products to be grown.

5. Giving control of the genetic structure of plants over to a few giants gives them huge power. Companies can copy and exploit genes from under the noses of the owners. For instance, the Neem tree in India has been used for medicines and pesticides for thousands of years. Now an American company has produced a pesticide from Neem tree genes. The Indian government says that the people of India should benefit from this invention; the company should pay India a fee for using their tree.

6. GM crops cannot be contained. Pollen from modified crops will spread to other plants nearby, and any harmful side effects will spread as well.

7. Dolly the cloned sheep shows us that cloning is possible. Another genetic engineering company is already cloning pets for their doting owners. Surely human cloning is inevitable unless it is made illegal worldwide.

8. Look at some of the crops that the GM giants choose to develop. A lot of research is going into tobacco, an addictive drug and a known killer, but a big earner for the companies who grow it. No research is going into cassava or millet, which are staple crops in many poor African countries.

9. Because of the excitement over GM foods, research into biological controls is being reduced. Biological controls are naturally occurring control techniques. For instance, 10 years ago 200 million Africans were threatened by a mealy bug that was eating the cassava crops. Hans Herren found a parasitic wasp in Paraguay that loved mealy bugs. He brought some over to Africa and the crops were saved. The problem for Hans is getting people to invest in this form of research. The wasps cost the farmers nothing, so there was no profit for Hans. Investors don't like that!

10. The fury that erupted among British shoppers in 1999 happened when people realised that they had been eating genetically modified soya for some time without knowing it. American laws did not insist that GM crops be labelled, so GM and non-GM soya were routinely mixed together. The shock of being so casually treated by the GM companies created anger around the world.

11. How long will the 'clever crops' last? Many genetic modifications result in crops that repel pests because they taste horrid, or even poison the insect. But pests evolve – eventually a resistant strain of insect will arise – and the problem returns.

12. Although the promise of GM crops is that they will remove the need for poisonous crop sprays, this hasn't happened. Farmers use even more sprays, knowing that their new GM crops can't be harmed by them.

What is life?

A few years ago, in 1996, it was announced that life had been found on Mars. Well, not exactly on Mars, but inside a meteorite that had come from Mars. And not exactly life, but tiny bacteria-like structures that had died millennia before. Meteorite ALH84011, said the discoverers, contained evidence that there could have been some form of primitive life on Mars millions of years ago.

As a theory, it lacks the buzz of 'little green men' popping out of craters and surfing down the Martian canals on their flying saucers. But even so, some said this was proof enough that Martians existed, even if they were only one hundred millionths of a millimetre long. Others poo-pooed it. Rubbish, they said, pure science fiction.

Life: unique or universal?

At times like this you find that scientists divide into two 'gangs', lined up against each other: those who think life is a miracle confined to this planet only, and those who are sure there are other life forms out there in space, waiting to be discovered. One lot say that it's impossible for the harsh conditions of Mars – or anywhere else for that matter – to support life of any kind. The other gang believes that some kind of primitive life is possible even in impossible conditions. They point to areas on Earth where conditions are harsh, so harsh that, until recently, nobody thought anything could survive there – deep within the arctic permafrost or in hot, underwater volcanoes or embedded in rocks kilometres underground. All these places used to be considered too cold, hot, poisonous or unreachable, yet some forms of bacteria have managed to set up home in them.

And what strange forms of bacteria they are! Much smaller than was ever thought possible – some only 20 millionths of a millimetre wide. When a thing is that small it's difficult to know whether it's really alive at all. That has raised the big question: 'What is life?'

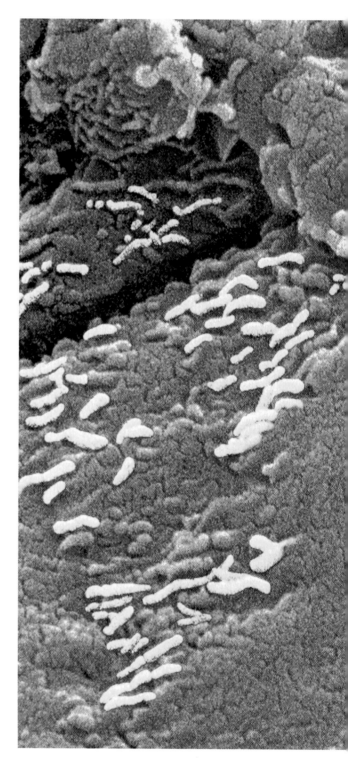

Evidence for life on Mars? Coloured electron micrograph of rods (coloured yellow) on a meteorite which came from Mars, discovered in Antarctica in 1984. Some scientists interpreted these as possibly being fossils of bacteria-like organisms.

In the beginning ...

The meteorite threw up another clutch of questions, even bigger ones, such as 'How did life begin on Earth?' The standard story goes as follows: the Earth formed about five billion (five thousand million) years ago. For the next billion years it was a seething mass of molten elements, a pungent soup in which atoms combined into molecules and split apart again randomly in a continuous 'molecular rave'.

One of the slightly more exotic molecules was able to do a neat trick – instead of breaking apart it could make an identical copy of itself. So while other molecules melted away into the soup, these kept increasing in numbers. Many scientists say that this was the start of life on Earth.

By about four billion years ago these molecules, now recognisable as the substance we call DNA, had evolved a protective coating, and were able to take molecules – nutrients, you could call them – from their surroundings. We know them today as bacteria. (So bacteria have been around for about 4000 million years. Humans appeared three million years ago. That's just minutes ago in geological time!)

The primordial soup - a mix of simple chemicals in which life on Earth may have begun.

The Martian 'seed'

Now, if this thing in meteorite ALH84011 turns out to be a Martian bacterium, that suggests another possibility, that life on Earth was 'seeded' by a meteorite from Mars. If there was life on Mars, what happened to it? Satellite pictures suggest there was water there once, enough to carve out river valleys, and there may have been the right conditions for life to begin. Something then caused the planet to lose its water and freeze to -100 °C. What was it? Could it happen here?

Perhaps by studying Mars we can help prevent the same tragedy befalling us here on Earth.

QUESTIONS

Scientists arguing over meteorite ALH84011 have come up with a number of definitions of life:

Living things have a regular structure

Living things move

Living things grow

Living things consume materials from their surroundings

Living things excrete waste products

Living things reproduce

Living things respond to environmental change

Living things contain DNA

With these definitions in mind, are the following things alive or non-living? Give your reasons for and against.

a) A robot

b) A pot of yoghurt

c) A salt crystal

d) A flame

e) BSE (mad-cow disease) is thought to be caused by a prion (a particle that acts like a virus, but has no DNA). Is a prion alive?

Mr Blobby

In 1994 some fishermen in the Baltic Sea spotted a curious, jelly-like blob swimming past their boat. It was huge, the size of a donkey, but unlike anything they had come across before; no eyes, no mouth, no features of any kind, but a definite sense of purpose judging by the way it was swimming towards the shore. Scientists examined it under the microscope – it was made up of trillions of identical amoebae, one-celled animals that normally live alone, but had somehow gathered together to form this creature. The scientists realised they had found the largest slime mould in the world!

Slime moulds are actually very common, though not usually that large. You may find them on rotting logs in gardens, as tiny patches of yellow slimy material. It's only recently that their extraordinary lifestyle has come to light.

An amazing catch! Not a fish but a huge marine slime mould!

Slime mould story

The story of your garden slime mould (*Dictyostelium discoistelium*) begins with a single spore floating in the air. The spore is very small and very light. It has been blown many kilometres before landing on a soggy log, perhaps, in your garden. When it cracks open the amoeba inside slithers out and straight away starts to eat, sucking at the wood and chewing up any bacteria it comes across. As it eats it swells. Soon it divides into two separate amoebae. The second amoeba starts to suck away just like mum, and soon they both divide again, into four, then eight, then sixteen ...

After a while the log is covered with millions of the microscopic blobs, all feasting away and producing ever more baby amoebae. Eventually, the end must come. That nutritious log has been sucked clean, the banquet is over, and now the army of amoebae face starvation. And it's then that something quite spooky happens. All those separate amoebae, as if by magic, turn inward and start to swim towards the centre to join up with the others into one great big glob. In their hour of need those individuals become part of a single entity, an animal that begins to move slug-like up the log. It has a front end that searches for the best route, and it moves in a co-ordinated rhythm.

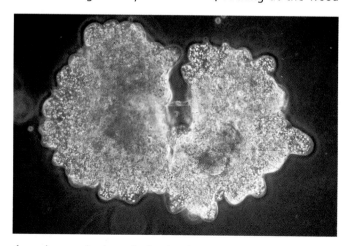

Ameoba – a single-celled animal.

Microscopic view of slime moulds feeding on decaying log.

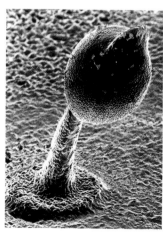

The escape pod forms.

Incredible morphing power

f the slime mould 'slug' is cut up into little pieces, each ndividual piece reforms as a miniature slug. If a little slug encounters another, they join up to form one larger beast.

When the slug finds a suitable spot an even more remarkable thing happens – it morphs again. The front begins to reform as a thin stem with a sphere on top, an escape pod in which one or two amoebae sit while the skin of the sphere hardens.

There it sits until a passing breeze wafts it away. The rest of the colony is doomed, but by a remarkable piece of teamwork they have given themselves the chance of a new life. This escape pod is able to travel far away to find a new home.

Clues to human origin

Scientists studying this transformation have been fascinated by what it might tell us about our origins. Three billion years ago our ancestors were amoebae. Did they at some stage find that living together in colonies helped them survive and prosper? Perhaps in these colonies different amoebae took on different jobs – one sort would specialise in finding food, others would deal with locomotion, and others would get on with making offspring.

When a human baby is just beginning, when it is only a small ball of cells, those cells all look the same; like the slime mould amoebae. They contain in their chromosomes detailed instructions for a whole human. At some stage, in some way that we still don't properly understand, sections of the chromosomes are switched on to make the cells specialise as bone, skin, muscle and so on, and the task begins of making the immensely complex lump of stuff that is a human being.

It is fascinating to watch the same thing happening there in a tiny corner of the garden. Perhaps if we study our garden slime mould a little harder we will get some more clues about how we are made the way we are.

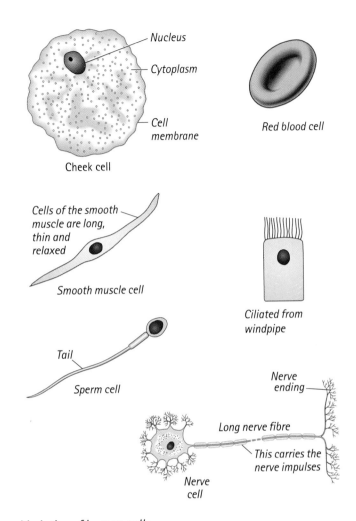

Varieties of human cells.

QUESTIONS

1. Draw a diagram of the amoeba cell in the photograph and label the nucleus, cytoplasm and cell membrane.
2. How does the 'slug' of a slime mould differ from a real slug?
3. The author describes the reproduction of the amoeba-like cells. If the cells divide every 30 minutes, how many would there be in 12 hours from a single cell?
4. How does the biology of the slime mould give some clues as to the way complex organisms might have evolved?

Aristotle's egg

'Another sunny morning. Living here in Mytilene is so different. So provincial. I cannot get so concerned about my appearance these days but it is difficult to get the latest fashions and my beard needs another trim. I find the young so uncouth in their manners. Perhaps I should stop complaining and say a little bit about myself.

I'm called Aristotle. I was born in Stagia, Asia Minor. What you now call Macedonia. My parents died when I was a boy and relatives brought me up. While my father was alive he taught me something of his work as a doctor. When I was 18 I went to Athens to study under the great thinker Plato.'

Plato talking to Aristotle, while other students at the School of Athens look on.

After Plato

'After Plato's death it was difficult being a Macedonian and following Plato's ideas, so I moved to Assos, a small port in Atarneus. Atarneus is in Ionia, in what is now Turkey. The ruler Hermisas had attracted many of Plato's students as his advisors. This made it possible to continue Plato's work. It was here I met my wife Pythias, adopted niece of Hermisas and we had our own daughter Pythias. Sadly, my dear wife died in childbirth.

Three years working in Atarneus and still the memories were difficult, so we moved to Mytilene here in Lesbos. That is where you find me now. I am trying to write up my observations of the hen's egg in my book *Historia Animalium*. I am distracted because I have to make a decision about becoming the tutor to Alexander, the son of Philip, King of Macedonia.'

Mytilene, Aristotle's home on the Greek island of Lesbos.

Miracle of the egg

'"Why watch a hen's egg?" I was asked the other day, by my colleague Theophrastus. I believe we can find nothing out about the workings of the human body without studying simpler animals. We have walked in the water around Lesbos and studied many of the sea creatures. Talking with fishermen has revealed much about the way marine life behaves. Cutting up and looking inside those creatures has told us a lot about the lives of sea animals.

Aristotle was a keen observer of living things, especially in the rock pools around Lesbos.

I also believe that all living things have a purpose and function in the cycle of life. Every organism has a destiny and shape related to its purpose. So small crabs have flattened rear feet to help them swim. They do not have tails like lobsters because they would be of no use in the shallow waters. But I go off the point.

The hen's egg, yes. I noticed that each generation of young from different organisms is identical to the parents, but the developing stages are different. The hen is a good subject since we can break gently into the shell and watch what happens beneath the membrane. Watching the hen's egg I have noted that the embryo forms after three days and nights. However, for bigger birds this period is longer and for smaller birds this period is shorter. Is there a relationship between the bird's size and the time for embryo development? Or is this just too few examples to reach that conclusion?'

Unanswered questions

'Meanwhile, the yolk rises to the sharp end with the embryo. It is important to note this is where the hen's egg breaks when it is ready for hatching. Is that because it is the weakest part of the egg? Next in the cycle is the development of the heart – red and clear. It soon starts beating and from it grow two tubes, like veins or ducts, into a membrane carrying thread-like blood vessels. Is the heart one of the essential features of life? Why does it develop so early on? And does the red blood fluid have an important role in the development of other organs?

Soon the body forms, with a clear part that can be seen to be the head and brain, with eyes that bulge and stare out at you. Could this mean they do not grow very much more after they have formed? The small embryo seems to take nourishment from the jelly-like white of the egg. How does it get this nourishment? It seems to be by way of a thin string that extends from the embryo in the yolk to the white. Why is there still a small space at one end of the egg? Is it there in all the eggs looked at? Yes it is, so it must have some purpose, but what is its purpose?

Ten days have passed. The chick is formed but the eyes have collapsed. They are less bulged but still large, more normal in appearance, with what looks like a hard black ball inside. We removed these eyes from a dead embryo and examined them more carefully. They are black and about the size of beans. We peel off the upper surface, like skin. Inside, the liquid is white and reflects light very well,

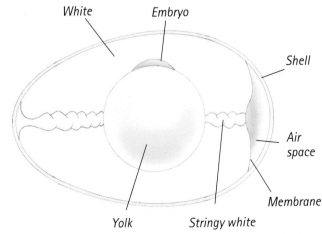

Aristotle made detailed observations of hens' eggs to find out how the embryo develops.

39

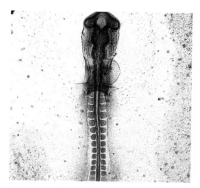

Chick embryo after 48 hours of development (head at top).

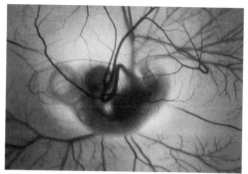

Living chick embryo, supported by network of blood vessels in egg.

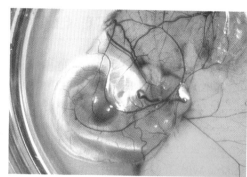

Head of chick embryo, showing large developing eye.

almost glittering in the bright sunlight. There is no hard ball in the centre of the eye. Looking more closely at the whole embryo we see the internal organs are formed. Does this mean they can work at this time? Is the chick embryo functioning as a living animal or a collection of organs?

Looking closely at the embryo's insides, or viscera, we can see the tiny blood vessels close to the stomach and very close to the navel and the string-like tube. There is also a vein stretching away from the embryo to the membrane surrounding the chick and the jelly-like white, which is now less in quantity and gathered at one end. This membrane keeps the chick separate from the liquid. There is a second membrane surrounding the whole collection. How does the chick breathe? Does it need to breathe?'

Final stages

'It is now twenty days and the chick can move inside the egg and can make chirp-like sounds. It is beginning to be covered in down. The head is resting close to the right leg and the wing is placed over the head. Inside the stomach we find yolk deposits, which indicates the young chick is taking nourishment from the yolk and this must be by way of the thin string. There is still a membrane surrounding the yolk, which seems to contain fluid that has been deposited by the chick from its stomach.

Shortly after the twentieth day the chick pecks its way out of the shell. It is interesting how the animal changes over the time of development. We really must look at the development of different birds and see if there is any similarity. I wonder, does the same development take place in a fish egg? Is there a space in the fish egg like that in the chick egg?'

QUESTIONS

1. The article discusses events that took place between 343-342BC. Find the dates of Aristotle's birth and death. Did Aristotle take up the job of tutor to Alexander?
2. Draw up a time line for change in the hen's egg. Can you think of a reason for that pattern?
3. What does the sequence tell you about the development of the living processes in an embryo? Work out your own explanation for this order and discuss the possible explanations.
4. Take some of the unanswered questions posed by Aristotle and see if you can research answers to those questions. Write your answers in the form of a letter to Aristotle.
5. Using the information on the following page, draw a time line from 500BC to AD500, showing when the named scientists lived. Include on your time line the ideas and work they are known for. Is there any pattern to their work or ideas?
6. AD450 is seen by many to be the start of the Dark Ages in Europe. In terms of science, try to give some explanation for this idea and try to find out what was happening in other cultures around the world.

Artemisia of Halicarnassus *Ionia* (Turkey) *c.*-490.
Navigator and sea captain.
Fought at battle of Marathon.

Cleopatra of Alexandria (Egypt) *c.*-450.
One of the first to find out about the human reproductive process.

Agnodike (Greek) *c.*-420.
Disguised herself as a man to learn medicine. Practised as a doctor until discovered. Sentenced to death for deception. Women protested so much the law prohibiting women from being doctors was changed to allow them to practise on women.

Elephantis (Greek) *c.*-75.
Doctor/midwife who used drugs from plants to help people with infertility and induce abortions.

Olympias of Thebes *c.*-40.
Doctor who wrote on the use of plant drugs as cures for infertility and inducing abortion.

Mary (Maria or Miriam)
the Jewess from Alexandria (*Egypt*) *c.*130 CE.
Alchemist who invented many pieces of laboratory apparatus.

Ban Zhao (China) 45-115 CE.
Alchemist. With her brother she wrote a book on 'Goldmaking' (Chrysopeia).

Theosobeia of Alexandria (*Egypt*) *c.*310 CE.
With her brother Zosimus she wrote an encyclopedia of alchemy.

Hypathia of Alexandria (*Egypt*) *c.*370–415 CE.
A mathematician who wrote an algebra book and invented the apparatus for distillation. She also worked with
Synesius of Cyrene, and with him invented the hydrometer for measuring density of wine and a silver astrolabe for finding position at sea.

Fabiola of Rome *c.*390 CE.
Founded one of the largest and most well-known hospices for the sick and needy. These hospices became the forerunner of the modern hospital. Many more were founded after this by religious orders. Hospices had been established in India for a number of centuries.

Anaxagoras of Clazomenae *c.*-500–428.
Explained that the Sun is a huge ball of red-hot stone. It is much larger than the Earth. The Moon reflects the Sun's light. He also stated a large number of 'seeds' make the properties of materials.

Empedocles of Akragas (*Sicily*) *c.*-490–430.
Put forward the idea there are four elements: earth, water, air, fire. Space is always in a state of confusion.

Democritus of Abdera, *Thrace* (*Greece*) *c.*-460–370.
Expanded his teacher, Leucippus', idea of atoms as tiny bits of matter. He showed how every form of matter could be explained using this idea.

Hippocrates of Cos (*Greece*) *c.*-460–377.
Wrote on scientific medicine and founded the profession of physicians, separating medicine from religion and superstition.

Strato of Lampsacus (*Turkey*) *c.*-340.
Conducted experiments to show falling bodies accelerate but incorrectly stated heavy bodies fall faster than light ones. He also worked on the lever.

Shih Shen, Gan De and Wu Xien *c.*-300–291.
Chinese astronomers independently developed star maps, which are used for centuries by astronomers and navigators.

Liu Hsin *c.*9 CE.
Chinese mathematician. First person to use a decimal fraction.

Galen of Pergamum (*Turkey*) *c.*130–190 CE.
Compiled a medical encyclopedia in which he put forward the idea of using the pulse as a tool for diagnosing illness.

Aryabhatta *c.*497 CE.
Indian astronomer calculated the measurements of the solar system and put forward the idea that the Earth rotates on its own axis. In 499 he also calculated pi – the ratio of a circle's circumference to its diameter – as 3.1416, 10 years after Chinese mathematicians Chu'ung-Chih and Tsu Keng-Chih calculated it as 3.14159203.

Bhaskara 1114–1185 CE.
Indian astronomer measured the diameter of the Sun.

(CE=Common Era). Same as AD.

(- = Before 0 in the Gregorian calendar. Same as BC).

What's this I see?

Sometimes, when we try to describe an event or an object to a friend, it can be hard to find just the right words to describe what we have seen.

Scientists often write descriptions of things or events they have observed. They try to be as accurate as possible in their descriptions. This is important because other scientists may use the descriptions in their own investigations. When describing an animal, for example, it is important to include as much information as possible. This will allow a reader to identify the animal if he or she finds one.

Medieval menagerie

In the Middle Ages, the early scientists wrote books describing all known animals and plants. Many of the animals and plants they described were used for making medicines. The descriptions had to be good enough for other doctors and scientists to identify which animal or plant was to be used to make a medicine.

The descriptions they wrote were not always very accurate. The writer had not always seen the animals and plants they wrote about! Their descriptions included details that had been passed on by others. Sometimes, the descriptions included evidence from stories told about the animal that were not accurate at all. Medieval books about animals even contained descriptions of unicorns and dragons!

When scientists make new discoveries they have to describe things that have never been seen before. When this happens they are often lost for words. When the scientist Antony van Leeuwenhoek first looked at drops of pond water through the first microscopes he could not believe what he saw. He was the first to observe microscopic animals and named them 'animalcules', which means 'tiny animals'. He even believed that he saw tiny human beings under the microscope, and added human faces to some of the animalcules he drew from his observations.

DESCRIPTION 1

This beast breeds in the River Nile. It is an animal with four feet, amphibious, generally about thirty feet long, armed with horrible teeth and claws. So great is the hardness of its skin that no blow can hurt it, not even if hefty stones are bounced on its back. It lies in the water by night, on land by day.

It incubates its eggs in the earth. The male and female take turns. Certain fishes which have a saw-like dorsal fin destroy it, ripping up the tender parts of its belly. Moreover, alone among animals it moves its upper jaw, keeping the lower one quite motionless. Its dung provides an ointment which old and wrinkled women use to anoint their figures and are made beautiful.

DESCRIPTION 2

I discovered little creatures in rain water which had been standing in a tub. Of the first sort I discovered that their bodies consisted of 5, 6, 7, or 8 very clear globules. I could not see any membrane or skin which held the globules together. These animalcules sometimes stuck out one or two little horns which continually moved, after the fashion of horses' ears. The part between these horns was flat. Otherwise their bodies were roundish, and ran to a point at the hind end. At this point it had a tail, near four times as long as the whole body, and looking as thick as a spider's web, when viewed through my microscope. At the end of this tail there was a pellet of the bigness of one of the globules of the body.

These little animals were the most wretched creatures that I have ever seen.

Animals in stories

Scientists are not the only writers who try to describe living things accurately. Many stories and poems contain beautiful and exciting descriptions of animals. The writers include details that make the animal more interesting to the reader. Although these descriptions are often as accurate as those written by scientists, this is not always so. For example, the rabbits in *Watership Down*, a story by Richard Adams, could talk!

DESCRIPTION 3

Gasterosteus aculeatus, 6-10 cm. This species has two large dorsal spines and a third small spine immediately before the single dorsal fin. The pelvic fins have a strong spine which can be locked open. There are no scales but bony plates along at least part of the lateral midline.

The animal feeds on small invertebrates. Breeding occurs in April and May, when red breasted males build a nest of plant material glued together with kidney secretion and defend the surrounding area. After a courtship ritual spawning takes place and the male guards and fans the eggs for 1–2 weeks.

DESCRIPTION 4

Huge it was – gigantic – standing on its hind legs more than twice as high as a man. Its shaggy feet carried great, curved claws as thick as a man's fingers. The mouth gaped open, a steaming pit of white stakes. The muzzle was thrust forward, sniffing, while the bloodshot eyes peered short-sightedly over the unfamiliar ground below. For a long moment it remained erect, breathing heavily and growling. Then it sank clumsily upon all fours, pushed into the undergrowth, the round claws scraping against the stones – for they could not be retracted – and smashed its way down the slope towards the red rock.

QUESTIONS

Read these four descriptions of animals.

1. Identify which description fits each of these animals:
    - crocodile
    - bear
    - stickleback
    - microscopic animal
2. What information in each account did you use to decide what the animal was?
3. What clues in the use of language in description 2, tell you that this was the earliest piece of writing?
4. Which writer is describing the first ever description of a newly discovered animal? What clues in the writing tell you this?
5. Which of the descriptions do you think come from a novel? What clues in the writing style tell you this?
6. Which description is the most 'scientific'? What clues in the writing style tell you this?
7. Choose an animal you are familiar with. You could choose a pet or a favourite animal from a film or book.
a) Write an accurate scientific description of the animal.
b) Write an account in a more informal style, describing what the animal is like. You could do this as a poem if you like.

Hello, Dolly!

Non-identical twins

I expect you know quite a lot about twins. You know that twins are exactly alike. Well, that isn't true. Many twins are different. One may be a boy and one a girl; you can't get more different than that! Or they may be the same sex, both girls or both boys, but look quite different. That is because they develop quite separately from two eggs, each one fertilised by a different sperm. So they are just like brothers or sisters, but born at almost exactly the same time.

Look at diagram 1 and you will see how the very first cell of each of these so-called 'non-identical' twins is made. It gets half its genes from mum and half from dad. The egg and the sperm are special cells with only half the usual number of genes.

Diagram 1. Non-identical twins develop from two eggs fertilised by two sperm cells.

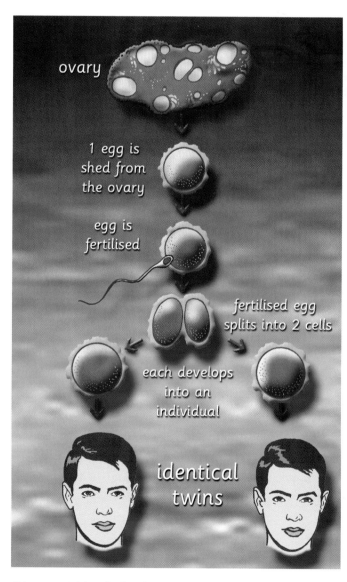

Diagram 2. Identical twins are produced when one fertilised egg divides into two and each half develops separately.

Identical twins

But when identical twins are born they look exactly the same because both have developed from the same fertilised egg cell. When twins are born the mother often asks, 'Are they identical twins?' Well, that is difficult for the nurses or the doctor to answer. They may look a bit alike; many babies do. The best way to decide is to see if there was only one amniotic sac. Identical twins grow in the womb side-by-side in one sac, which is sometimes a bit of a squash. No one knows why the egg divides into two before the two babies start growing. No one knows either why having identical twins 'runs in the family'.

Look at diagram 2. The identical twins both come from the very first cell. They are so similar that even the parents have some trouble telling one from the other at the beginning. Even the babies get muddled about who they are! You could call them clones.

Plant propagation

Sometimes, plants make a new plant when the two sex cells combine, just like animals do. The egg comes from the ovary at the base of the flower, and the male cell comes from the pollen. The seed may grow in the ground and make a new flower. It might be a different colour or have differently shaped petals because the genes from the egg and from the pollen have combined that way.

Look at diagram 3. The new plant is definitely not a clone of its parents, and each one of the new plants may be different.

Plant cloning

But you can make new identical plants quite easily. If you cut off a good piece of a plant, dip its end into a hormone mixture and then plant it in the ground at the right time of

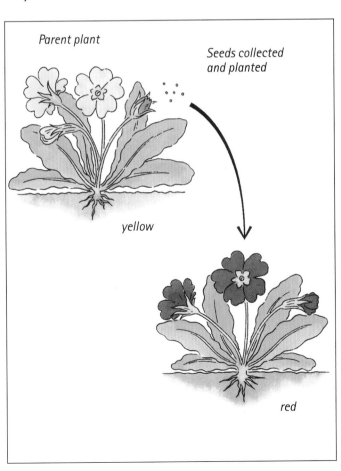

Parent plant

Seeds collected and planted

yellow

red

Diagram 3. A primrose that grows from seed may not look exactly like its parent.

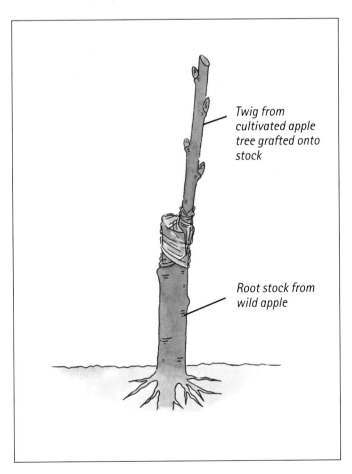

Twig from cultivated apple tree grafted onto stock

Root stock from wild apple

Diagram 4. Fruit trees are usually grown by grafting a twig of the fruiting variety onto the root of a different variety. **45**

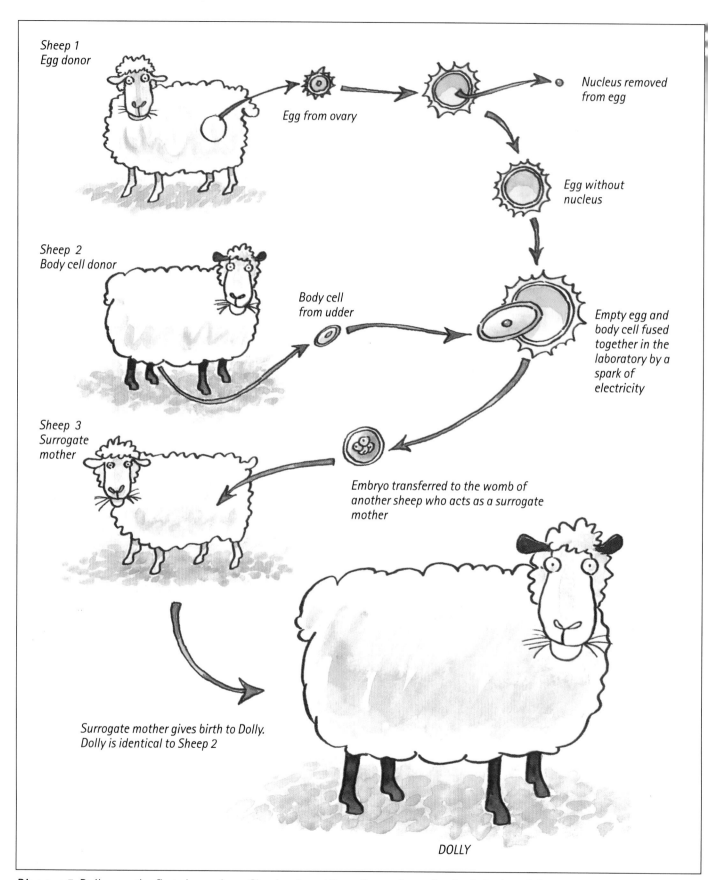

Sheep 1
Egg donor

Egg from ovary

Nucleus removed from egg

Egg without nucleus

Sheep 2
Body cell donor

Body cell from udder

Empty egg and body cell fused together in the laboratory by a spark of electricity

Sheep 3
Surrogate mother

Embryo transferred to the womb of another sheep who acts as a surrogate mother

Surrogate mother gives birth to Dolly. Dolly is identical to Sheep 2

DOLLY

Diagram 5. Dolly was the first sheep clone. She developed from the nucleus of a single body cell from her black-faced mother, which was made to fuse with the empty egg cell from a white-faced sheep.

year, it will often grow. Then the new cells will be just like the old cells. It's called a cutting. Every cutting from a plant will be a clone of the original. It does not have two parents.

Look at diagram 4. Fruit trees are usually made like this. If you take an apple and plant its pips the new trees will make new kinds of apple. To get a Cox's Orange Pippin apple tree from another one you always need to take a cutting. But you can tell from its name – 'pippin' – that the very first one was grown from an apple pip.

New cells for old

Can we make an animal clone from just one parent? Most cells grow inside the baby, but skin cells only grow into skin, and muscle cells into more muscle. So the scientists have to take an egg cell from the ovary, which might grow into a whole baby. However, this has only half the normal number of genes in its nucleus. So that nucleus is removed, and a complete nucleus from an udder cell is put in its place. Then the new egg cell is put back in the womb. Sometimes it does grow all the way into a new animal. 'Dolly the Sheep' was made in this way.

Look at diagram 5. See if you can follow all the stages in making Dolly, a clone of her mother. Isn't it amazing? She has grown up to be exactly like her mother. If her mother needed a transplant, they could use body tissue from Dolly, without fear of rejection.

Scientific breakthrough – at a cost

That was a scientific breakthrough! Some farmers and scientists thought that the next stage might be to clone the best cows, sheep and pigs in much the same way. But it's not that easy! When they first tried to make Dolly the scientists cloned 277 eggs, taking out the nucleus and then putting in a complete nucleus from the mother. Only 29 of them even began to develop into normal embryos. Dolly was the only one that became a normal baby lamb and did not die soon after birth. There have now been some cloned calves, but only four out of 10 embryos survived. Mice are much easier to clone.

Why is cloning so unreliable and the cloned animals so weak? Scientists are making hypotheses and will try out many more experiments to improve the cloning technique. Their hypotheses include:

- Is it because the cloned cells are really much older than they seem?
- Is it because the clones still have some strange different genes from the egg cell?
- Is it because the scientists have to use an electric shock to make the new nucleus begin growing in the empty egg cell?

The scientists' work is not finished. There is a lot more research to be done.

QUESTIONS

1. Where are the genes in a normal cell?
2. Farmers usually improve their cows by mating a very good milking cow with a bull. But it does not always produce very good calves. Why is that?
3. How would you clone a very good milking cow?
4. Which of the possible reasons why cloned animals so often die do you think might be right?
5. Some genes exist inside cells but they are outside the nucleus, in structures called 'mitochondria'. These genes are called 'mitochondrial genes' and they are made of 'mitochondrial DNA'. Look up those terms on the Internet and see what you can find out about them.

Dinosaur eggs

In 1923, Roy Andrews, an explorer and expert on fossils, led a team of scientists on an expedition to a red sandstone area of the Gobi desert in Mongolia. The expedition was looking for evidence of the earliest human fossils, which they expected to find in this area. What they did find turned out to be even more exciting!

The area they explored was hundreds of kilometres from the nearest inhabitants. The explorers were living in tents in the middle of a desert more than ten times the size of Great Britain.

Egg-citing discovery!

One morning, while out searching for good fossil sites in the desert, Roy stumbled across a complete nest of fossilised eggs. The eggs were too large to have been laid by any known bird, past or present. In fact they turned out to be the first fossilised dinosaur eggs to be discovered. On a previous expedition, part of an eggshell had been found, but this latest chance discovery was more than anyone expected.

The eggs were studied closely back at the scientists' base. Evidence showed that they had been laid about 80 000 000 years ago, towards the end of the Cretaceous Period. They were found near the surface, partly exposed from the sand by millions of years of erosion.

Roy Andrews discovered a nest of fossilised dinosaur eggs in the Gobi desert.

The first dinosaur remains

Scientists had discovered dinosaur remains for the first time in 1833. The early discoveries in the United States and Mexico excited many people by revealing that a totally unknown group of animals – the dinosaurs – had dominated the world millions of years before humans came along. Since dinosaurs are extinct, scientists can only find out about their lives by studying the clues left in the fossil remains. Palaeontologists – the scientists who study fossils – had often wondered how these great beasts reproduced. Here at last was startling proof that at least some dinosaurs laid eggs. Not only that, this particular dinosaur laid its eggs in a nest, much as turtles or birds lay their eggs today.

Further searching at the site provided clues as to the identity of the dinosaur that had laid the eggs. Fossilised bones found near the nest were carefully pieced together. Complete skeletons were found of a dinosaur called *Protoceratops.*

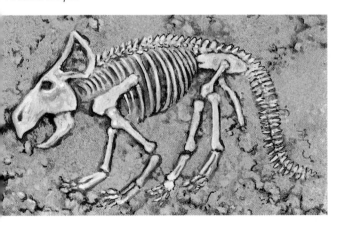

A complete fossilised Protoceratops *skeleton was found near the nest.*

Skeletons of both adults and young *Protoceratops* were found, suggesting that the animal lived in family groups in which the parent looked after their young to some extent. A fully grown *Protoceratops* was about six metres long, and could be recognised by its gigantic beak. These were the forerunners of the great horned dinosaurs, including the three-horned *Triceratops.* On the back of its head was a large bony frill to which the jaw and head muscles were attached. It also protected the otherwise vulnerable neck.

Conditions for fossil-making

The brick-red rocks in which the 15 eggs were found was made from fine grains of red sand that had been pressed together. The rocks were soft and crumbly and were probably formed by sand blown by the wind. Mongolia must have been a very dry, hot desert when *Protoceratops* was alive. Sand storms, much like those in deserts today, would have made living conditions very difficult at times.

The eggs may have been laid in the nest just before such a giant sand storm. Covered too deeply by the drifting sand, the heat of the Sun's rays could not reach them to help the embryo dinosaurs grow. They became embedded more and more deeply in the sand. Over many years the pressure of the great weight of the sand above them caused the sand around them to become compressed to form rock. During this time the eggs themselves had been replaced by sand and so became fossilised.

Conditions in Mongolia were ideal for forming fossils. There would have been very little water, and air was kept from the remains by drifting sand. In these conditions the remains of the dead animals could not decay. And because little of Mongolia has been submerged beneath the seas since *Protoceratops* became extinct, very little of the old sandstone in which the fossils were found has been covered by sedimentary rock (rock formed underneath water). Many fossils can be found near the surface.

Life in the desert

So *Protoceratops* lived in desert conditions, possibly near to local water sources such as streams and ponds. Its horny beak and the shape of its few teeth suggest that it was a plant eater. It probably wandered the desert around the water hole, plucking leaves and branches off the desert shrubs.

Groups of these giant reptiles probably moved into the desert to find safe places to lay their eggs. They would search for suitable sand dunes and prepare a round pit for the females to lay their eggs. The eggs were laid in circles with the fat ends pointing towards the middle of the nest.

The fossil evidence suggests that female Protoceratops
*fought to protect their eggs from carnivorous dinosaurs
such as* Oviraptor.

Each had a hard shell for protection. The herd of
Protoceratops probably stayed around the pits to guard the
eggs from predators.

   An enemy of the *Protoceratops* was a large ostrich-like
dinosaur, *Oviraptor,* (which means 'egg stealer'). Part of the
skeleton of one of these beasts was found on top of a nest
of *Protoceratops* eggs at the site. It is likely that this
creature was actually in the act of stealing the eggs when
the sand storm blew up and buried it along with them!

   Roy Andrews went on to make many more explorations
in Alaska, the East Indies, China, Korea and Mongolia. He
was made head of the American Museum of Natural
History in New York in 1935.

QUESTIONS

1. Scientists use clues in the fossils to help them
   make informed guesses about how the animals
   lived. Read carefully through the account of
   the discovery of *Protoceratops* eggs and
   answer these questions:

   a) What clues suggested that the eggs were
   laid by *Protoceratops*?

   b) What clues suggested that *Protoceratops*
   fed on plaɪts?

   c) What clues suggested that *Protoceratops* lived
   in groups that looked after their young?

   d) What clues suggested that *Oviraptor* fed on
   the eggs of other dinosaurs?

2. Draw a series of diagrams to show how the eggs
   of *Protoceratops* may have become fossilised.

3. Imagine that you have travelled back in time to
   observe life on Earth. You find a round nest
   containing large eggs. Nearby are some strange-
   looking dinosaurs, feeding on plants. A large
   ostrich-like dinosaur approaches the nest.
   Write a description of the events that follow.

4. Imagine that you are Roy Andrews. You are very
   excited to be the first to discover dinosaur eggs.
   Write a letter to a friend explaining how you
   feel and how important the discovery is.

Magnetic animals

Every year thousands of Bewick's swans arrive at wetland areas on the east coast of Britain. The swans have just completed a two-month journey that started 4000 kilometres (2500 miles) away on their summer breeding grounds in arctic Russia. They will spend the winter in Britain where the climate is warmer and there is plenty of food. Then in spring they will return to Russia to raise their young. Each year the Bewick's fly the same route and stop off at the same wetland feeding areas.

Millions of birds migrate in response to the changing seasons. Birds migrate to avoid harsh weather, to find food and to find safe breeding grounds. The longest migration journey is made by the arctic tern, which flies from the far north of Europe and North America to spend the winter along the shores of the Antarctic. This is a distance of over 20000 kilometres (11000 miles). The round trip means that the birds spend about eight months on migration.

How are birds able to fly such huge distances without getting lost? How do they know in which direction to fly? It has been known for centuries that birds have a very good sense of direction, but scientists are only now beginning to understand how birds know where they are going.

Secrets of navigation

Birds use a number of different methods to navigate (find their way). 'Piloting' is when birds follow a series of recognisable landmarks along their migration route. These landmarks are memorised over repeated journeys. Bewick's swans travel around the coastline of northern Europe as they migrate from Russia to Britain. It is likely that they recognise rivers, mountains and features of the coastline along the route. The birds will use these features to remember where to find good stopover places. Young swans will learn the route from their parents.

Many birds have to migrate across hundreds of kilometres of sea, where there are no landmarks. These birds cannot use piloting to find their way. Pacific golden plovers migrate from Canada to the Polynesian islands, that is 4000 kilometres (2500 miles) without land! These birds know what direction they are travelling in by the position of the Sun. Because the Sun moves from east to west during the day, birds can only use it if they know what time of day it is. Fortunately, birds have an internal body clock that tracks the daily light-dark cycle. Therefore they can calculate their direction based on their internal clock and the Sun's position. For example, if flying north, they know that the Sun should be kept on their right in the morning and their left in the afternoon.

Bewick's swan in flight.

Scientists think that birds also sense patterns of light in the sky. And it is these patterns of light that birds use to navigate rather than the position of the Sun. The Earth's atmosphere scatters sunlight into patterns. The patterns of the scattered, or 'polarised', light change as the position of the Sun in the sky changes. If you learn the patterns associated with different positions of the Sun, you can work out where the Sun is, even if it is behind clouds. To be able to see these light patterns you must have very special eyes. Humans don't have these special eyes, but birds and bees do. These light patterns are very clear just when the Sun is setting. Lots of birds migrate during the night, therefore they may use the light patterns at sunset as a cue to start flying.

Night flyers

The majority of birds migrate at night. When birds fly at night they use stars to find their way. In the northern hemisphere birds use the North Star because it is particularly bright in the sky. Birds are able to remember the patterns of the stars in the sky. Scientists used birds that had never seen a real sky, to show that birds do use the stars to navigate. The roof of a planetarium was painted so that it matched the positions of the stars in the sky. When birds were let inside, they started to fly in the same direction as if they were migrating. Experiments have shown that birds are able to navigate with just partial glimpses of the stars. This is important on cloudy nights.

Human navigators use the position of the Sun as a rough estimate of direction. In order to be more accurate we use a compass and the Earth's magnetic field; the magnetic field that surrounds the Earth and causes compass needles to point north. Birds are able to sense the Earth's magnetic field. Scientists think that birds as well as some insects, fish, sharks, turtles and mammals are able to navigate over long distances using the Earth's magnetic field. Even termites are able to sense the Earth's magnetic field – they build their mounds facing east-west. If a termite mound faced south it would absorb the hot midday sun and the termites inside would overheat.

The internal compass

To be able to sense the Earth's magnetic field, these animals must contain an internal compass. Up until recently scientists did not know what this internal compass was made of or how it worked. But experiments with fish and ants are beginning to show how animals sense the Earth's magnetic field. Scientists working with trout have discovered tiny particles of magnetite in the trouts' noses. Magnetite, an oxide of iron, was the first mineral to be used by humans to make magnetic compasses more than a 1000 years ago. Magnetite has also been found in animals

ranging from bacteria, insects, sharks, turtles and rats to humans! In ants, magnetite was found in the head and body of the animal. The ant uses its whole body as a compass needle and can sense its heading relative to the magnetic field.

How are animals able to read their internal compasses? And how is this information transmitted to their brains? These are two questions that are still a mystery to science. In trout, the cells containing magnetite particles were found very close to a network of nerve cells. These nerve cells must be involved in telling the brain what direction the internal compass is facing. In birds, magnetite particles have been found very close to the optic nerve. The optic nerve is a dense cluster of nerve fibres behind the eye that transmit light patterns to the brain. Scientists think that, at least in birds, the ability to sense the Earth's magnetic field is related to their sight. If the birds can also detect the strength of the Earth's magnetic field, they may even be able to tell how close they are to the north and south poles (the Earth's magnetic field gets stronger towards the poles).

Pigeon experiments

The discovery of magnetite particles in birds has started a rather strange set of experiments. In these experiments homing pigeons are fitted with magnets on the front of their heads. Homing pigeons are known for their ability to find their way back to their roost over hundreds of kilometres. The results showed that these birds find it difficult but not impossible to work out where their home is. This may mean that there is more than one built-in compass needle to be found in birds!

Scientists still have a long way to go before they understand completely how animals navigate. Birds use a range of methods to help them find their way. If one method doesn't work, say if the sky were cloudy, then the bird is able to use its compass instead. By using all these methods of navigation, birds are less likely to get lost.

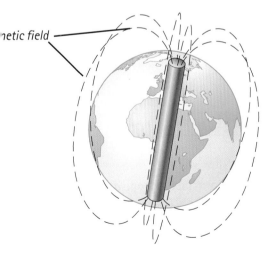

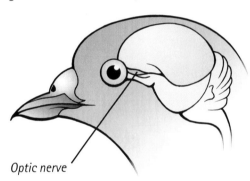

Optic nerve

netic field

Particles of magnetite – iron oxide – have been found next to the optic nerves of pigeons. The birds may be able to detect the Earth's magnetic field to help them navigate.

QUESTIONS

1. Why do birds migrate?
2. The Bewick's swans migrate from arctic Russia to Britain. They stop off in Matsalu Bay, Estonia, Ringkobing Fjord in Denmark and the Lauwersmeer in Holland. On a map of Europe trace the route taken by the swans. Identify features that they might use to find their way.
3. Why do you think that birds migrate at night and not during the day?
4. Why do scientists use fish to study migration behaviour, while there have been relatively few experiments with birds?
5. Many scientists wonder whether humans have an in-built sense of direction. Why don't you try an experiment in the classroom to find out if people can navigate? You need a chair that swivels, a blindfold and some willing volunteers.

Can we learn to love wolves?

What do you think about fox hunting? Are you for it, or against it? Foxes are shot by farmers if they start attacking their chickens or lambs. There are plenty of foxes about and no amount of hunting or shooting is likely to wipe them all out. Foxes are not an endangered species, indeed they are on the increase with many of them learning how to live in towns by feeding on food thrown away in dustbins.

The situation of elephants and tigers in Africa and Asia is quite different. These animals are rare. They are endangered species and it is strictly against the law to shoot them, even if they attack adults or children. They may be terrorising a whole village but permission has to be given to kill them. How would you feel about this if your baby brother had been killed by a lion or an elephant? You might be very angry if some westerners wanted to protect the killers, even if they were very rare.

Return of the grey wolf

In Europe our most ferocious killer, the grey wolf, has been an endangered animal and protected by law since 1979. Now it seems to be coming back! The last wolf in England was killed in the eighteenth century, and a hundred years later they were wiped out in France, Switzerland and Germany. There have always been plenty of wolves in eastern Europe, deep in the forests of Poland and Russia, and now they are spreading west and increasing in numbers. Does that frighten you? Every European country has its stories about 'big bad wolves', just as we do. It is an ancient native animal of Europe, and has probably lived here as long as people. Most scientists believe that all domesticated dogs are descended from the wolf.

The male adult wolf is about 2 metres long, including its 1 metre of bushy tail. That is as tall as a tall man. It is a very intelligent animal and lives in packs of about 20 wolves of all ages. There is a dominant pair, the father and mother, and the others have all been their pups at different times. They usually have litters of six to seven in April. You may have seen dogs marking out their territory where other dogs are not allowed to enter. Wolves do the same. A wolf pack may claim hundreds of square kilometres of land. They then defend this territory against all others, and howl loudly to tell other packs not to come near. This means they need a lot of land.

Grey wolves live and hunt in packs.

What's on the wolf menu?

What do they eat? How will they manage in our populated countries? Wolves do attack sheep. If the sheep huddle together, as 'silly' sheep tend to do, the wolves find them easy to kill. They may kill far more than they really need, just as foxes kill more chickens than they can eat. (Probably these animals have an in-built memory of famine and being hungry.) In a pack they can also attack cows and horses, although this is very rare. Many European farmers want to shoot all wolves on sight, but there are other ways of protecting sheep. One way is to raise sheep who do not naturally huddle together when frightened, and the other is to protect the sheep with sheepdogs. No doubt some would get eaten, especially young lambs in the spring. The territory of wolf packs is so large that some farmyards and fields are almost bound to be included.

Do wolves attack people? The answer is 'almost never'. There are terrible stories from long ago of wolves attacking unprotected babies during very severe winters when they were very, very hungry. But they could not get into our homes to do that now.

What was the place of the wolf in the natural food chain when it used to live in England? It always came at the very top. It would eat almost any herbivore (grass-eater), like rabbits and deer. Sometimes the pack would attack foxes just as our foxhounds do, and then eat them. All of those animals are very common in England now. Some people reckon that four rabbits eat as much grass as a sheep, and there are an awful lot of rabbits around! Rabbits are mostly nocturnal animals (coming out at night), just like the wolves do. Farmers find it very hard to control the numbers of rabbits on their farms, and some of them might even welcome a little help from wolves!

The wolf – friend or enemy?

Deer always used to be the favourite food of wolves. A strong stag (male deer) can run away from a wolf pack, or even charge them with his antlers, but the old and sick animals are less likely to escape. These are the animals that might spread disease. At present the gamekeepers who look after the deer have to shoot the old animals, and also some of the younger ones, to make sure that there are not too many for the amount of food. (The gamekeepers are acting as predators – like the wolves used to be.) Too large a herd of deer will strip the leaves and growing shoots from the trees, which destroys them. Not enough grass may make the deer hungry. They are very good jumpers and can often leap out of the woods into the neighbouring farmers' fields, if they smell food there. If half their crop of wheat is eaten, the farmers will not be too pleased with the deer either!

Stag by Loch Ossian, Scotland.

Can European people live with a new population of wolves? Maybe the animals will learn to live on the edge of cities eating our waste food – discarded hamburgers, meat and hot dogs. Foxes do that. In Italy, where wolf numbers are increasing, they have already been seen strolling round the piazzas (city squares) at night without causing trouble. But most farmers want to shoot them all dead. Perhaps, if the government gives them compensation (money) for any sheep killed, the farmers might feel better. Once upon a time wolves were a part of our habitat, although not much loved. One French environmental officer said about the war between farmers and wolves:

"Forty dead sheep are not a reason for opening fire."
Do you agree?

QUESTIONS

1. Draw some food chains showing how energy from the Sun feeds the wolves.
2. Describe two ways in which the wolf might help the farmer.
3. Why do people in Poland and Russia sometimes hear wolves howling at night?
4. Answer the last question in the article, giving your reasons for and against.

Jane Goodall

Children's toys can be very important to them. When Jane Goodall was one year old her mother gave her a large, hairy toy chimpanzee. It had been made to celebrate the birth of a chimpanzee at London Zoo. Most of the friends of Jane Goodall's mother were horrified. They predicted that the 'ghastly creature' would give the child nightmares. However, Jubilee, as she was called, became Jane's favourite toy.

Jubilee may have been important not only for Jane Goodall but for the world of science too. For she has spent her life studying chimpanzees in the wild. Her work has completely changed how we see chimpanzees and has led to new ways of studying animal behaviour.

By the time she was eight years old, Jane Goodall had decided that when she grew up she would go to Africa and live with wild animals. She left school at 18 and trained as a secretary. While she was working she received an invitation from a school friend to visit her at her parents' farm in Kenya. Jane Goodall handed in her resignation the same day and soon left for Africa.

A chance meeting

Out in Africa she happened to meet a famous scientist, Louis Leakey. They talked together about chimpanzees. Leakey told Jane that it was time someone started a two-year study of chimpanzees in the wild. As they talked, Jane guessed what was coming. Yet she could scarcely believe her ears when he suggested that she should carry out the study! After all, she had never even studied animal behaviour let alone gone to university to do a degree in the subject.

As Jane Goodall herself has written:

'Louis, however, knew exactly what he was doing. Not only did he feel that a university training was unnecessary, but even that in some ways it might have been disadvantageous. He wanted someone with a mind uncluttered and unbiased by theory ...'

Beginning the work

When Jane Goodall arrived to start her study in 1960, the game warden who took her round made a mental note that she wouldn't last more than six weeks. She has stayed over forty years and is there to this day.

Dr Jane Goodall is a British ethologist – a scientist who studies animal behaviour – and chimpanzee specialist.

At first she found it impossible to get close to the animals. They simply disappeared whenever she came near. Then, one magical day, some six months into her study, Jane stumbled upon two male chimpanzees, David Greybeard and Goliath. Scarcely breathing, she waited for them to run, as they always did. But they didn't. Less than 20 metres from where she stood, they started to groom one another. Soon, all the animals accepted her and she was able to watch them and make notes in their presence without affecting their behaviour.

Is it science?

One of the really important things about Jane Goodall's work was that she carried it out in the wild. Now, this means that she couldn't do the sort of experiments that many scientists do. She couldn't carry out a 'fair test'. She couldn't control variables. And she couldn't take repeated sets of measurements.

Another interesting thing about Jane's work was that she gave the animals names. She thought of them as individuals, each with their own personality. At the time, this was very unusual. Scientists weren't used to thinking of animals like that.

Indeed, when she came to interpret her findings, Jane Goodall didn't hesitate to talk about chimpanzees 'getting restless', 'wanting to go places', 'setting off resolutely', 'flying into a tantrum' and 'pottering about'. She believes that chimpanzees spend a lot of time thinking. They think about ways of getting food, ways of getting to mate with particular individuals, and of how to keep their place in the hierarchy.

Nowadays, most scientists agree with Jane Goodall. She has changed the way animal behaviour is studied and she has changed the way we see animals.

Are chimpanzees like us?

Jane Goodall's work on wild chimpanzees has led her to campaign against chimpanzees being kept in small cages and used for medical research. In the UK this is no longer allowed though most countries still permit it.

The reason chimpanzees are used in medical experiments is that they are very like us in many ways. This means that we can learn a lot about human drugs and vaccines by testing them on chimpanzees. But the fact that chimpanzees are very like us makes many people think that it is wrong to experiment on them.

Something special about women

Jane Goodall is, of course, a woman. Interestingly, long-running studies on animal behaviour are usually carried out by women. Jane Goodall has worked on chimpanzees from 1960 to the present day. Dian Fossey worked on gorillas from 1966 to 1985, when she was murdered, probably because of her work. Fiona Guinness has worked on red deer from 1972 to the present day.

QUESTIONS

1. Why did Louis Leakey want Jane Goodall to study chimpanzees rather than someone who had studied animal behaviour at university?
2. Why do you think the chimpanzees gradually accepted Jane Goodall and stopped running away from her?
3. Do you think the fact that Jane Goodall couldn't carry out any experiments means that her work is not good science? Explain your answer.
4. List arguments for and against using chimpanzees for medical experiments.
5. Can you think of some possible reasons why long-term studies on animal behaviour are often carried out by women?
6. Imagine you had the chance to study the behaviour of one animal species for several years. Which animal would you choose and why?

Controlling the fox

At about one metre in length, the fox is Britain's largest natural predator. Foxes hunt alone, mostly at night, when they seek out small prey. They feed mainly on rodents, such as field mice, and other small mammals of the woodland and countryside. They also eat fruit and nuts at times, especially in the autumn. During the day, foxes tend to stay under cover.

Fox lifestyle

Parent foxes live in an underground den called an earth, which they dig out among the roots of large trees. Sometimes they take over holes dug by other animals, including badgers. In the spring they produce litters of up to ten cubs, but five is more common. The parents take care of the young, bringing food back to the den each night to feed them. A hungry litter can get through a great number of rats, voles, rabbits and other animals.

After about one month, the cubs come out from the earth to play and explore their environment. They stay with the mother for protection, leaving her in the autumn to make their own lives, and fend for themselves.

The diagram of a food web on the next page shows some of the feeding relationships that foxes are involved in. One of the fox's traditional food sources was the rabbit. In the past, people would trap rabbits and sell them for food. The rabbit trappers were in competition with the fox and often set traps for it. But during the 1950s a disease called myxomatosis wiped out much of the rabbit population. With the decline in the number of rabbits, many of the rabbit trappers stopped their work. As a result the fox was left to breed with little control of its numbers.

One of the more cruel ways of trapping foxes – with gin traps – was made illegal in the 1950s, so fox numbers continued to rise.

Red fox with pups, next to their den.

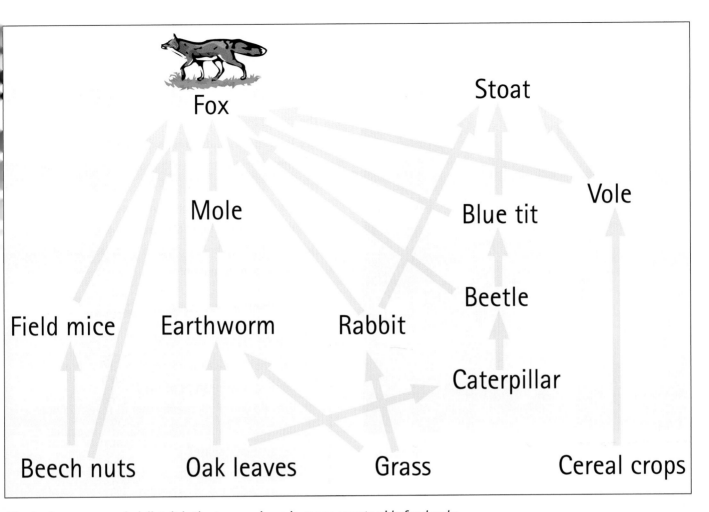

Fox Stoat Mole Blue tit Vole Beetle Field mice Earthworm Rabbit Caterpillar Beech nuts Oak leaves Grass Cereal crops

The fox has a very varied diet. It is the top carnivore in many countryside food webs.

New food sources needed

With a good food supply foxes can breed rapidly. They take care of their young, so many survive to increase the numbers in the population. There are many more foxes now than forty years ago, therefore they have had to look for new sources of food.

Foxes are well known for their cunning and versatility, and have adapted well to man-made changes in the environment. They can cope well in cities and even in industrial areas. Early morning commuters often see foxes along the railway embankments and on factory sites.

As well as wild creatures, foxes may kill hens, newborn lambs, and ground-nesting birds. Unfortunately, they have a habit of 'surplus' killing, which means the fox kills many more animals than it needs for food, and so there are occasions when the fox population must be controlled to prevent damage.

The fox hunt

Humans are the fox's only natural enemy. Many people believe that it is important to kill foxes to stop them from attacking farm animals for food. One way in which humans try to keep the number of foxes down is by hunting. Fox hunting is one of the most familiar forms of hunting in England. Packs of hounds can be found in many parts of the English countryside. Boxing Day is the most popular day for hunting, and at that time many thousands of people are out taking part in, or watching, the hunt.

The hunters, with their pack of foxhounds, join the hunt followers at a pre-arranged time and place to begin the hunt. The hounds have been trained to search for foxes. They are encouraged to search for the scent of the fox. When they smell a fox's scent, they follow its trail. The riders chase after the hounds. They are skilled horsemen, and must jump high hedges to keep up with the hounds.

A huntmaster leads a pack of foxhounds while on horseback during a foxhunt.

If the hounds catch the fox they attack and kill it almost immediately. Sometimes the fox goes to earth, before it is caught. It hides beneath the ground in an animal's burrow. The hounds can't get to it. When this happens, small dogs called terriers are sent into the burrow. They flush the fox from its earth. The fox may then be given a start before the hunt is resumed. Sometimes it's killed with a humane killer.

Many hunts look after large areas of the countryside to improve their hunting. This makes an attractive landscape, and also provides many habitats for wildlife in addition to the fox. They believe that hunting is the best way to control the numbers of foxes living in an area. At the same time they enjoy the pleasure gained from the thrill of the chase.

Stop the hunt!

But many people in Britain think that fox hunting is a cruel sport and should be banned. They believe that the hunters are barbaric people who get their enjoyment from seeing the fox savagely killed.

Many of these people try to stop fox hunts from happening. They try to help the fox by getting in the way of the huntsmen. Sometimes they lay false trails to confuse the foxhounds when they are searching for the scent. Some of the protesters believe that foxes should simply be left alone. They argue that farmers should find other ways of protecting their animals, which do not involve being cruel to wildlife.

Others believe that if foxes are a pest to a farmer, then their numbers should be controlled in a less cruel way. They could be culled by being shot so they are killed cleanly, with as little pain as possible.

QUESTIONS

1. Use the food web diagram to describe the feeding relationships of the fox.
2. Why do you think the number of foxes did not go down when the rabbit population declined because of disease?
3. Suggest three reasons why the number of foxes has increased so much over the last 40 years.
4. Some people think that foxes are becoming a pest. Suggest three reasons why they think this.
5. What do you think would be the consequences if fox hunting was made illegal?
6. **Read the statements about fox hunting on the next page:**
a) Decide which people are in favour of fox hunting and which are against it.
b) Write a list of five points in favour and five points against fox hunting. You may use different ideas from those shown here.
c) Work with two partners to write a script for a ten-minute radio show about the issues to do with hunting. You could play the roles of an interviewer, a person who is for and one who is against hunting.
   Act out the script.

Adam

'In a hunt, the fox is either killed or it gets away. A hunted fox is never left injured.'

Bill

'Foxes that get away from the hunt are more wary – the barking of farmyard dogs might keep them away from hens and lambs.'

Emma

'Fox hunting provides many jobs for people who live in the countryside.'

Colin

'Foxes that are caught are often the old, injured or weak. This leaves the healthiest ones to continue breeding.'

Dave

'The hounds are large and strong compared with the fox, and therefore kill it instantly.'

Elvis

'When the hounds kill a fox it is very quick and painless.'

Manny

'Hunting may injure the fox and cause it unnecessary pain, even if it gets away.'

George

'Many of those who hunt do so for the thrill and excitement of the ride. They enjoy themselves even if they do not find a fox.'

Fred

'People who follow the hunt must be cruel and barbaric to enjoy seeing a fox ripped to pieces.'

Harry

'Hunting gives the fox a chance to escape. It is less cruel than putting out a trap or shooting it.'

Ian

'Using so many hounds to chase just one fox is very unfair to the fox.'

Anna

'Shooting the fox would be a more humane way to kill it because it wouldn't know what had happened.'

Liam

'Using poison, snares and traps may cause a lot of suffering to the fox, and may injure other species that are not harmful.'

Neil

'There is not enough evidence to prove that foxes are a real threat to lambs and other domestic animals.'

Robert

'The hunt helps to preserve the countryside for other uses such as tourism.'

Peter

'Hunting frightens other animals in the countryside.'

Chris

'Chasing foxes over the countryside causes a lot of damage to farms and crops.'

Triffids

In John Wyndham's novel *The Day of the Triffids* there were large man-eating plants that roamed the land capturing and devouring any unfortunate person that happened to cross their path. Is this possible? What could plants possibly do to us?

First, plants are wonderful makers of powerful poisons. They can sting us with thorns to inject these compounds so that we itch, swell up, become paralysed or even die. Why should they do this? Some plants do it just to protect themselves, like brambles or cactuses. In north America there is a common plant called 'poison ivy', which can make people so ill that they can die. But there are some plants that really want to kill animals so that they can eat them! Are you frightened yet? Don't worry, they eat mostly insects.

Plants could easily adapt to hearing us since they often have fine hairs on their stems. If these were just a little more protected they might be able to vibrate, like the cilia inside our ears. When our cilia move they pass messages down nerves to the brain so that we can hear. Do you like the idea of plants listening out for you as you move through a field or a park in the dark? Are you frightened yet?

No problem, you think? We could fire off guns at any hostile plants. They could not fire back, could they? But have you seen the pink Himalayan balsam, which grows wild all down the river banks in southern England. In the summer, as you walk down the river path they can shoot you at a distance of ten metres. Only seeds? Well, yes.

Ancient myths and legends suggest that there were once man-eating plants that grew in the remote jungles of distant lands. Perhaps they just hid and waited for someone to pass by. Could they see? There don't seem to be any modern plants that can do that, although many react to light. Some South African geraniums actually fold up their leaves into little parcels when the sunlight is too bright. It is weird, and rather sweet to watch them do that.

Triffids - carnivorous plants with a taste for human prey!

But in John Wyndham's story about triffids, the plants were strong, 3 metres tall, and had three legs. They could see well enough to lash out at a person's head. Now that is really very frightening! Fortunately, remember, triffids don't exist.

Carnivorous plants – the facts

The truth is that there are over 400 known species of plants that can trap and digest small animals, particularly insects. These are insectivorous or carnivorous plants.

The plants live in soils that have very little of the essential element nitrogen. Most plants get this from rotting material in the soil, from fertiliser or from nitrogen in the air. But these plants obtain the nitrogen by digesting the bodies of small animals and absorbing their rotting remains directly into their leaves.

The Venus flytrap

The best known example of an insectivorous plant is the Venus flytrap, which is found only in North and South Carolina, in the United States. The leaves are specially adapted to form traps. Insects are attracted by the smell of nectar and the bright red colour of the inside of the trap. A single trap of this plant is divided into two lobes with spikes along their edges to stop the insect escaping when the trap snaps shut. They really do snap shut! There are six small trigger hairs inside each trap, three on each side. One hair touched twice, or two hairs touched once, signals the trap to close over the unfortunate victim. Digestive enzymes, produced by the plant, digest the soft parts of the insect into a kind of liquid fertiliser. Empty traps reopen in a day or so, but traps with prey remain closed for a week or two. Then, the trap will reopen, exposing nothing but the dry shell of the victim.

Venus flytrap - leaf closing over insect.

The pitcher plants

There are many different species of pitcher plants found mainly in the tropical parts of the world. The largest of them grow like vines and may be up to 15 metres tall. The pitcher plants get their name because the leaves curl upward to form jug or pitcher-shaped tubes. Although most of the pitchers capture small insects, the largest of them may devour large beetles, frogs, snails and even small birds and mammals. The bottom of the pitcher is filled with a mixture of rainwater into which the plant secretes enzymes. Insects are attracted to the pitcher by its colourful stripes and the nectar produced by nectar glands on the inside lip of the pitcher. As the insect crawls closer to the edge of the pitcher, it loses its footing on the slippery, downward-sloping hairs and falls into the liquid. Enzymes or bacteria begin working almost immediately, while the stiff hairs prevent the escape of the unlucky insect.

Pitcher plant.

The sundew – living fly-paper!

This plant lives in the boggy areas of moorland such as Dartmoor. It gets its name from the tiny drops of dew-like secretion produced by the leaves, which may then be seen to glisten in the sun. Like other carnivorous plants the leaves are adapted for the business of trapping insects. In the sundew the small leaves are arranged in a rosette close to the ground. Each leaf is covered in tiny hairs that produce a sweet-smelling sticky liquid to attract the insect. Any insect landing on the leaf will become stuck on the hairs, which then bend over and entrap the insect even

Sundew.

Bladderwort with trapped mosquito larva.

more securely. The hairs at the centre of the leaf then produce the enzymes, which begin to digest away the prey. So the sundew behaves like a living fly-paper.

The bladderworts

Most of these are rootless plants that live in ponds, although some of them have become adapted to living as air plants on the trunks of tropical trees. They have a spectacular way of trapping insects. The finely divided leaves contain tiny bladders, which act as traps. They are no more than 5mm in diameter. On the front of the bladder is a trapdoor, covered in sensitive hairs. As the small insect brushes against the hairs it causes the trapdoor to suddenly open and, in a fraction of a second, the insect is sucked into the bladder. Once inside the bladder, the trapdoor shuts and enzymes begin to digest the insect while it is still alive. Although most of the prey is made up of small water animals such as mosquito larvae and water fleas, the largest of the species may have bladders big enough to ingest a small tadpole. As the walls of the bladder are translucent, the unlucky prisoner can be seen

thrashing about helplessly inside the trap! It takes a period of about 15-30 minutes for the trap to reset, and several days for the prey to be digested.

QUESTIONS

1. Which organ of the Venus flytrap has been adapted to trap insects?
2. What would be the normal function of this organ?
3. Explain why these plants have to use insects as their source of nitrogen.
4. It is possible to buy these plants to grow at home. Why do the instructions tell you 'DO NOT FEED WITH FERTILISER'?
5. Imagine you were asked to grow several Sundew plants in the laboratory. Describe how you would set up a container for growing them. (You might like to search the Internet for websites on carnivorous plants – some of these will give you growing instructions.)